# PROTEIN FOLDING-MISFOLDING: SOME CURRENT CONCEPTS OF PROTEIN CHEMISTRY

# PROTEIN FOLDING-MISFOLDING: SOME CURRENT CONCEPTS OF PROTEIN CHEMISTRY

JOSEPH P. ZBILUT
THOMAS SCHEIBEL
EDITORS

Nova Science Publishers, Inc.
*New York*

For permission to use material from this book please contact us:
Telephone 631-231-7269; Fax 631-231-8175
Web Site: http://www.novapublishers.com



**Library of Congress Cataloging-in-Publication Data**

Protein folding-misfolding : some current concepts of protein chemistry / Joseph P. Zbilut and Thomas Scheibel, editors.
p. ; cm.
Includes bibliographical references and index.
ISBN 13 978-1-60021-417-2
ISBN 10 1-60021-417-7
1. Protein folding. I. Zbilut, Joseph P., 1948- II. Scheibel, Thomas.
[DNLM: 1. Protein Folding. 2. Proteins--chemistry. QU 55.9 P9664 2006]
QP551.P69582135 2006
572'.633--dc22 2006025999

*Published by Nova Science Publishers, Inc. ✣ New York*

# CONTENTS

# FOREWORD

Proteins occupy a unique position in the hierarchy of structural organization. The analysis of proteins is lying at the interface between physics, chemistry and biology. From the biological side, although any individual protein would not be considered to be alive, it does not take many of them (plus a bit of nucleic acid) before life-like behavior begins to emerge. While only specific sequences of amino acids are good for generating functional proteins, there are only weak departures of real protein sequences from random strings. For a chemist proteins are large complicated molecules that even polymer chemists would have difficulty in modeling. From a physical viewpoint, proteins are linear heteropolymers that, unlike most synthetic polymers, which are condensed from one or a few monomer units, are formed by basically non-periodic juxtaposition of 20 different monomers.

In contrast to typical chemical polymers which are generally very large extended molecules forming a matrix, the majority of proteins have the ability to self-assemble and to fold into relatively compact self-contained structures: each protein found in nature has a specific three-dimensional structure and this structure is determined by the sequence of the polypeptide chain. Such behavior leaves the particular linear arrangement of amino acids an efficient recipe for the solution of a chemico-physical problem that basically is the folding to a given unique three-dimensional protein structure in water solution.

And yet these peculiar amino acid strings are one basis of life. Although genomics has been the center of interest over the last decade, the fact that there are arguments for the primacy of proteins has been little appreciated. Strikingly, there is evidence that proteins and metabolism came before genes in pre-biotic evolution. Further, it is known that that protein association represents an essential problem in biomedicine and biotechnology. Particularly, a number of human disorders, including the several neurodegenerative diseases originate from the deposition of misfolded protein aggregates. In each of these pathological states, a specific protein or protein fragment switches from its natural conformation into an insoluble one which can accumulate in a variety of organs and tissues.

Clearly, the relationship between sequence-embedded information and folding behavior of proteins has been a dominant concern of biochemical research. What has been of significant interest is the lack of principles connecting the microscopic world of atoms with the macroscopic world of proteins. While we do have an accurate knowledge of potentials (hydrophobic interactions, hydrogen bonding, size constraints, etc.) acting at microscopic levels, the "mesoscopic" principles needed to predict the 3D structure of proteins remain essentially unknown. The crucial role of hydrophobicity in modulating folding and stability features of proteins has been firmly established by a large variety of approaches. Yet, a more detailed understanding remains elusive.

For this reason, protein research continues to be an intriguing area of research: the field spans the range from quantum to system, and requires some knowledge of not only the biological, but the physical sciences as well. Increasingly, familiarity with computational methods and statistics is becoming more important to sift through the mass of accumulated data.

Although, progress in this area is well attested by the numerous journals, books, and reviews extant, unfortunately, the beginning student is hard-pressed to find introductory publications that would provide a guide in this quickly evolving field. For this reason, the present monograph was conceived. Our aim has not been to provide an encompassing overview of the latest developments in protein science. Instead, we have sought to provide some outlines of basic themes in the field of protein folding, as well as some rudimentary expositions which can function as a basis for further exploration.

**Joseph P. Zbilut**
Department of Molecular Biophysics and Physiology
Rush University Medical Center, Chicago, IL USA

**Thomas Scheibel**
Lehrstuhl für Biotechnologie,
Technical University of Munich, Germany

*Acknowledgements*: JPZ would like to acknowledge the support from a joint DMS/DGMS initiative to support mathematical biology, from the National Science Foundation and National Institutes of Health, (NSF DMS #0240230).

# CONTRIBUTORS

Johannes Buchner, Technische Universität München, Department Chemie, Lehrstuhl Biotechnologie, Lichtenbergerst. 4, D-85747 Garching, Germany; johannes.buchner@ch.tum.de

A. Keith Dunker, Center for Computational Biology and Bioinformatics, Department of Biochemistry and Molecular Biology, Indiana University School of Medicine, Indianapolis, IN 46202, U.S.A. and Molecular Kinetics, Inc., Indianapolis, IN 46268, USA; kedunker@iupui.edu

Stephan Frey, Technische Universität München, Department Chemie, Lehrstuhl Biotechnologie, Lichtenbergerst. 4, D-85747 Garching, Germany; stephan.frey@ch.tum.de

Dirk Gillespie, Department of Molecular Biophysics and Physiology, Rush University Medical Center, 1653 W. Congress Pkwy., Chicago, IL 60612 USA; Dirk_Gillespie@rush.edu

Alessandro Giuliani, Istituto Superiore di Sanitá, Dipartimento Ambiente e Connessa Prevenzione Primaria, Viale Regina Elena 299, 00161 Roma, Italia; alessandro.giuliani@iss.it

Martin Haslbeck, Technische Universität München, Department Chemie, Lehrstuhl Biotechnologie, Lichtenbergerst. 4, D-85747 Garching, Germany; martin.haslbeck@ch.tum.de

Hauke Lile, Institut für Biotechnologie, Universität Halle, Kurt Mothes Strasse 3, D-06120, Halle, Germany; Hauke.Lilie@biochemtech.uni-halle.de

Sayam SenGupta, Technische Universität München, Department Chemie, Lehrstuhl Biotechnologie, Lichtenbergerst. 4, D-85747 Garching, Germany; sayam.gupta@fiberlab.de

Thomas Scheibel, Technische Universität München, Department Chemie, Lehrstuhl Biotechnologie, Lichtenbergerst. 4, D-85747 Garching, Germany; scheibel@ch.tum.de

Manfred Sippl, Center of Applied Molecular Engineering, University of Salzburg, Jakob Haringer Str.1, A-5020 Salzburg, Austria; sippl@came.sbg.ac.at

Vladimir N. Uversky, Center for Computational Biology and Bioinformatics, Department of Biochemistry and Molecular Biology, Indiana University School of Medicine, Indianapolis, IN 46202, U.S.A. and Molecular Kinetics, Inc., Indianapolis, IN 46268, USA and Institute for Biological Instrumentation, Russian Academy of Sciences, 142290 Pushchino, Moscow Region, Russia; vuversky@iupui.edu

Joseph P. Zbilut, Department of Molecular Biophysics and Physiology, Rush University Medical Center, 1653 W. Congress Pkwy., Chicago, IL 60612 USA; jzbilut@rush.edu

# PREFACE

Protein research continues to be an intriguing area of research: the field spans the range from quantum to system, and requires some knowledge of not only the biological, but the physical sciences as well. Increasingly, familiarity with computational methods and statistics is becoming more important and receiving more recognition in the masses of accumulated data. This new book provides outlines of basic themes in the field of protein folding, as well as some rudimentary expositions which can function as a basis for further exploration.

Protein folding describes the process of transforming linear information encoded by an amino acid sequence of a polypeptide chain into a defined three-dimensional structure. Although it has been established many years ago that protein folding is a spontaneous reaction, a detailed understanding of this fundamental process is still missing. In this context, the discovery of conformational diseases, which are based on complex conformational conversions of proteins accompanied by self-assembly, has added an additional twist into studying protein folding. Protein folding is often driven by a series of discrete individual steps. In many instances individual components do not contain all the information needed, but require guidance by "molecular chaperones" (however, such guidance may simply serve to preclude undesirable alternative self-assembly pathways, such as aggregation). Therefore, protein folding could be described as a highly convergent synthesis protocol that is often largely driven by non-covalent interactions. As a consequence, the same non-covalent interactions are responsible for preserving the integrity of the final protein structure. The complexity of the problem, together with the paucity of interpretable data, has generated a large volume of literature about protein folding, self-assembly mechanisms, and conformational conversion. Chapter 1 will provide basics on protein folding and will further give insights into trigger events that are crucial for self-assembly and conformational conversion.

Most soluble proteins are biologically active in protein complexes and oligomeric structures. In order to understand the function and stability of these complexes the molecular mechanisms of protein association and the structural features of the subunit interfaces need to be elucidated. Whereas statistical analyses of these interfaces reveal the importance of hydrophobic interactions within protein contact sites, a detailed analysis of several interfaces shows diverse strategies in accomplishing specific protein-protein interactions. The hydrophobic surface of the interface might be formed as a large hydrophobic core surrounded by polar groups or as the sum of small patches interspersed by polar residues and water molecules. In higher oilgomeric complexes one might find more hydrophobic and more hydrophilic contact sites at the different levels of association. In addition to the discussion of general aspects of protein-protein interactions, protein oligomerisation based on specific structural characteristics of a given protein such as domain swapping, coiled coil structures and four-helix bundles are discussed in Chapter 2.

Chapter 3 describes a family of intrinsically disordered proteins, members of which fail to form rigid 3-D structures under physiological conditions, either along their entire lengths or only in localized regions. Instead, these intriguing proteins/regions exist as dynamic ensembles where there are significant temporal variations in atom positions and backbone Ramachandran angles. Many of these intrinsically disordered proteins still carry out important biological functions originating from regions that lack specific 3-D structure. The existence of such proteins does not fit the structure-function paradigm, which dominates modern protein science and states that unique 3-D structure is a prerequisite to function. Thus, the classical structure-function paradigm has to be reconsidered to include intrinsically disordered proteins and alternative relationships between protein sequence, structure, and function. This shift in the paradigm represents a real breakthrough for biochemistry, biophysics and molecular biology, as it opens new levels in understanding of the complex life of proteins. This chapter will try to answer the following questions: How intrinsically disordered proteins were discovered? Why these proteins do not fold? What is so special about intrinsic disorder? What are the functional advantages of disordered proteins/regions? What is the functional repertoire of these proteins? What is the relationship between intrinsically disordered proteins and human diseases?

After sequencing of a substantial set of genomes is completed, one of the key problems of today's structural genomics and proteomics projects is high-throughput expression and purification of a large number of target proteins. As outlined in Chapter 4, the efficient production of many gene products is required in therapeutic research or for the decoding of the function of all proteins of one organisms. Traditionally, the production of a recombinant protein requires a cloning step of the target gene into an expression vector followed by its expression in bacteria, yeast or mammalian and insect cells. However, in many cases the recombinant proteins are not produced in substantial yields and in an active form. Thus, the development of efficient expression and purification systems is a major goal. While conventional microbial *in vivo* expression systems seem rather ill-suited for high throughput technologies, cell-free systems might be the tools of choice for the fast expression of proteins in parallel and for screening of a wide range of expression conditions. In addition, *in vitro* translation systems turned out to be very useful to study the influence of molecular chaperones on protein expression and folding.

Our ability to predict protein structures from amino acid sequences depends on our understanding of molecular forces. The same applies to the validation of protein structures determined in the laboratory. The protein structures available in the public domain contain a number of deficiencies and inconsistencies. As long as we are unable to recognize and correct such errors on a regular basis we cannot completely trust the experimental results and we cannot have any confidence in predicted structures. Chapter 5 investigates the theory of intermolecular forces from the perspective of protein structure theory, comment on the fundamental ideas involved, discuss the difficulties encountered, and we provide examples to illustrate the current state of affairs in protein structure validation and prediction.

As explained in Chapter 6, recurrence quantification analysis (RQA), is a signal analysis method to study sequence/structure relationships of proteins, a technique derived from nonlinear dynamics, which exposes and quantifies patterns in higher dimensional spaces. The basic focus of applications is in generating predictive sequence/structure or sequence/activity models on the basis of RQA hydrophobicity sequences of various protein families.

Since charges are ubiquitous in biology and electrical forces are very strong, an accurate physical understanding of a biophysical system requires an accurate and complete understanding of electrostatics. Chapter 7 describes the fundamentals of studying electrical charges and the forces acting on them through the electric field. Advanced topics of how charge distributions of mobile particles and the electrical potential are coupled are also discussed.

In: Protein Folding-Misfolding...
Editors: J.P. Zbilut and T. Scheibel, pp. 1-33

ISBN: 1-60021-417-7


*Chapter 1*

# FOLDING, SELF-ASSEMBLY AND CONFORMATIONAL SWITCHES OF PROTEINS

***Sayam SenGupta and Thomas Scheibel****

Technische Universität München, Department Chemie, Lehrstuhl Biotechnologie
Lichtenbergstr. 4, D-85747 Garching, Germany

## Abstract

Protein folding describes the process of transforming linear information encoded by an amino acid sequence of a polypeptide chain into a defined three-dimensional structure. Although it has been established many years ago that protein folding is a spontaneous reaction, a detailed understanding of this fundamental process is still missing. In this context, the discovery of conformational diseases, which are based on complex conformational conversions of proteins accompanied by self-assembly, has added an additional twist into studying protein folding. Protein folding is often driven by a series of discrete individual steps. In many instances individual components do not contain all the information needed, but require guidance by "molecular chaperones" (however, such guidance may simply serve to preclude undesirable alternative self-assembly pathways, such as aggregation). Therefore, protein folding could be described as a highly convergent synthesis protocol that is often largely driven by non-covalent interactions. As a consequence, the same non-covalent interactions are responsible for preserving the integrity of the final protein structure. The complexity of the problem, together with the paucity of interpretable data, has generated a large volume of literature about protein folding, self-assembly mechanisms, and conformational conversion. Our review will provide basics on protein folding and will further give insights into trigger events that are crucial for self-assembly and conformational conversion.

## I. Introduction

Proteins have diverse biological functions ranging from DNA replication, transporting $O_2$ within an organism to catalyzing various organic reactions [Creighton 1993; Whitford 2005].

* E-mail address: thomas.scheibel@ch.tum.de, Phone: +49-89-289-1 31 79, Fax: +49-89-289-1 33 45 (Corresponding author)

The types of functional properties of proteins are endless and the number of types keeps increasing as new protein functions are characterized constantly. Proteins have evolved through millions of years of selection and perform specific functions. The specificity of their functions is due to their unique three-dimensional structures. The three-dimensional structure arises because particular sequences in the amino acid chain assemble and fold to compact domains starting from a linear chain [Branden and Tooze 1991]. The folded domains can either provide a binding or catalytic site for a variety of functions or serve as modules for building up larger assemblies like virus particles or muscle fibers. The key elements and motifs of three-dimensional protein structures are described in greater detail within this introduction.

## A. Primary Structure

The primary structure of the protein is the linear sequence of amino acids along the polypeptide chain. The amino acids of the polypeptide chains are connected by amide bonds. Each protein has a unique sequence of primary amino acids. The information of the native three-dimensional structure is encoded in this primary amino acid sequence. Changes in primary sequence may or may not change the overall structure. If changes occur between closely related residues (e.g. leucine to isoleucine) the three-dimensional structure is conserved. Changes which are not conservative can lead to protein misfolding as is seen in various protein misfolding (= conformational) diseases (see V).

## B. Secondary Structure

A protein's secondary structure refers to the local conformation of amino acid residues. In globular proteins, there are predominantly three types of secondary structures: the α-helix, the β-sheet and the turns.

### 1. The α-helix

The α-helix is a classic element of protein structure. It was first theoretically predicted by Linus Pauling in 1951 [Pauling 1951] and then corroborated by diffraction patterns of proteins like hemoglobin. α-helices in proteins are found when all amino acids in a stretch of consecutive residues have a ψ, φ angle pair with approximately -60° and -50° (Figure 1). The typical α-helix has 3.6 residues per turn with hydrogen bonds between C=O of residue n and N-H of residue n+4. Different side chains have been found to have weak but definite preferences either for or against forming an α-helix. Thus Ala, Glu, Leu and Met are good α-helix formers, while Pro, Gly, Tyr and Ser are very poor. Such general tendencies of amino acids are good indicators towards prediction of secondary structures but it is important to mention that predictions are often inaccurate.

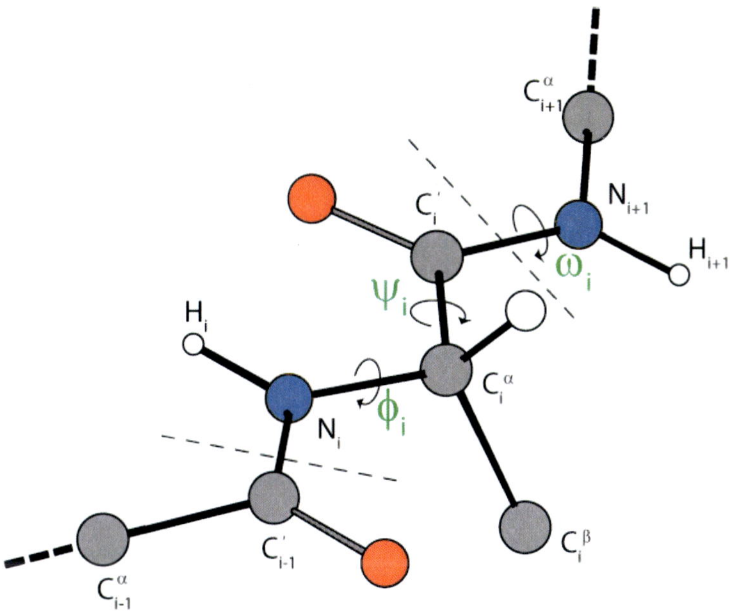


Figure 1. The perspective drawing of a segment of polypeptide chain showing the torsion angles ψ, ω and ϕ of the amino acid I, flanked by the amino acids i-1 and i+1. The polypeptide chain is shown in a fully extended conformation where ψ = ω = ϕ.= 180°. In an a-helical conformation ψ = -60° and ϕ = -50°.

### 2. The β-sheet

Unlike α-helices which are always built from one continuous amino acid stretch, β-sheets can also be built from a combination of several regions of the polypeptide chain. Single β-strands are usually 5-10 residues long and are aligned adjacent to each other such that hydrogen bonds can form between C=O groups of one β-strand and N-H groups on an adjacent β-strand and vice versa. β-strands can arrange next to each other to form sheets. The directionality of adjacent β-strands leads to the formation of 'parallel', 'antiparallel' and 'mixed' β-sheets.

### 3. Turns or Loops

Most protein structures based on combinations of α-helices and β sheets are connected by turns or loops of various lengths and irregular shapes. Turns have the universal role of enabling the polypeptide backbone to change direction and in some cases to reverse back on itself. Analysis of the amino acid composition in turns reveals that bulky or branched side chains occur at very low frequencies. Instead, residues with small side chains such as Gly, Asp, Asn, Ser, Cys and Pro are observed predominantly.

## C. Tertiary Structure

The tertiary structure represents the three dimensionally folded polypeptide chain. Elements of the secondary structural motifs interact and fold into a compact globular structure. These interactions include non-covalent bonds such as electrostatic interactions, van der Waals interactions, hydrophobic contacts and hydrogen bonds between non-backbone groups, as well as covalent bonds such as disulfide bridges.

*1) Charge-charge interactions*

These interactions occur between amino acids having ionizable side chains as in Lys, Arg, His, Asp, and Glu residues as well as between $NH_3^+$ and $COO^-$ groups at the termini of the polypeptide chains. Oppositely charged side chains (for example the $-NH_3^+$of Lys and $-COO^-$ of Glu) interact mostly by a combination of electrostatic forces and also with some degree of hydrogen bonding. These interactions are also known as salt bridges. The electrostatic interaction between two opposite charges A and B can be modeled by Coulomb's law as shown in equation 1.

Equation 1:

$$\Delta E = \frac{Z_A Z_B \varepsilon^2}{Dr_{ab}}$$

$Z_A$ and $Z_B$ are the number of charges on atoms A and B respectively, $\varepsilon$ is the charge of an electron, D is the dielectric constant of the solvent and $r_{ab}$ is the distance between the atoms A and B.

The stabilization energy, as seen above is inversely proportional to the dielectric constant. Therefore, charge-pain interactions between residues that lie on the surface of the protein are dramatically weaker than the ones which are in the interior due to the very high dielectric constant of water (D=80). Conversely, stabilization energy from charge-charge interactions between residues that are buried in the interior of proteins is much higher.

*2) Electrostatic interactions between dipoles*

Electrostatic interactions can occur between non-ionizable amino acid side chains because electron density can be localized if atoms have different electronegativities. Atoms with the greatest electronegativities have an excess of negative charge, δ-, others an excess of positive charge, δ+. For example, since O atoms are more electronegative than C atoms, C=O bonds of a peptide will have a net δ+ on the C atom and δ- on the O atom. This leads to the formation of dipoles. Two side-by-side dipoles repel each other when parallel and attract when they are oriented in an anti-parallel manner. However, in a three dimensional protein structure, the dipoles can be aligned in both parallel and anti-parallel orientation, decreasing the overall contribution of this interaction to the stability of the protein. In sense, they are weaker than those between ions (as seen in charge-charge interactions) because the net force in such an interaction is always attractive.

A classic example for this type of interaction is the π-π interaction between aromatic side chains. The π electrons in the aromatic rings of Phe, Tyr and Trp residues are localized above and below the face of the ring. This excess of electrons gives the face of the ring a small negative charge, whereas the hydrogen atoms on the edge have a slight positive charge. The π-π interaction originates from the electrostatic interaction between these partial charges. Aromatic rings preferentially interact with either the positively charged edge of one ring pointing to the charged face of the other or with their rings parallel but offset so that the edge of each ring is interacting with the face of the other.

*3) Hydrogen bonding*

A hydrogen bond is formed when the hydrogen atom covalently attached to a donor atom interacts favorably with an acceptor atom. The main component of the hydrogen bond is the interaction between the δ+ of the H atom bonded to donor and the δ- of the electronegative acceptor. Hydrogen bonds have huge impact on the formation of helices and β-strands in proteins. Importantly, hydrogen bonds can be formed between side chain groups and between main chains and side chains. The chemical groups on proteins that most commonly serve as hydrogen-bond donors are N-H, O-H, and, much less frequently, S-H and C-H groups. The most common acceptors are O=, -O-, -N=, and much less frequently, -S-, $-S^-$, and the π electrons of aromatic groups. The amino acids which are therefore typically involved in hydrogen bond formation are the side chains of Tyr, Thr and Ser containing the hydroxyl group and the side chains of Gln and Asn with the amide group. The strengths of hydrogen bonds are generally within the broad range of 2-10 kcal/mol. The strengths and lengths of hydrogen bonds depend on the electronegativities of the acceptor and donor; the greater their electronegativities, the shorter their lengths and the stronger the hydrogen bond.

*4) Van der Waals interaction*

These are interactions that arise due to induced polarization effects. In a simplistic model, a spherical atom may be considered to have a transient dipole arising from temporary asymmetry of the electrons and the nucleus. This transient dipole can polarize a neutral atom nearby and generate an attraction. These are very weak forces which are electrostatic in nature and vary as the sixth power of distance between them, $r_{ab}^{-6}$. The interaction between two induced dipoles is very important as it occurs among all atoms and molecules. There are both attractive and repulsive van der Waals forces that control interaction between atoms and are therefore very important in protein folding.

*5) The Hydrophobic effect*

During protein folding the amino acids with hydrophobic side chains (like Ala, Phe, Leu, etc.) tend to be buried in the interior of the protein to shield them from water. This phenomenon leads to a favorable free energy increase commonly known as the hydrophobic effect. The preference of amino acids with hydrophobic side chains for the interior of the protein, which resembles a non-aqueous environment, is due to hydrophobic interactions. Hydrophobic interactions are the strongest non-covalent intermolecular forces in water. In the tertiary structure of proteins, hydrophobic interactions lead to clusters of distal and proximal hydrophobic groups. The strength of such interaction is dependent on the 'hydrophobicity' of the interacting amino acid side chain. Hydrophobicity of an amino acid is defined as the free energy of its transfer from water to a non-polar solvent (like cyclohexane, octanol). This free energy of transfer is always favorable. In the unfolded state of the protein, the water molecules around a non-polar residue are more tightly packed than the bulk water. Upon folding, these water molecules are released leading to a gain in entropy. Hence the favorable transfer of a non-polar amino acid from water to a non-polar solvent is a result of the favorable change in entropy.

*6) Disulfide bridges*

Disulfide bridges dictate a protein fold by forming strong covalent links between cysteine side chains and are often widely separated in the primary sequence. As they are widely separated, the formation of the stable covalent disulfide linkage brings various secondary structure motifs near to one another. Disulfide bridges are mostly important for proteins that are secreted or involved in the secretory pathway.

### D. Quaternary Structure

Many proteins contain more than one polypeptide chain. The structure formed upon the interaction between single tertiary structures of more than one polypeptide chain (or subunit) is known as quaternary structure. The interactions responsible for the quaternary structure are the same as for tertiary structures. Quaternary structures can be based on identical subunits or non-identical subunits.

## II. Protein Folding: Self Assembly of Polypeptide Chains

Globular proteins are translated as linear chains of amino acids by the ribosome encoded by mRNA. However, for these proteins to function in biological processes, they need to adopt their specific three-dimensional structure. The process of folding from the unfolded state (linear chain of amino acids) to the folded state depends on the cooperative action of the relatively weak non-bonded interactions discussed above. The critical contributions to gaining the native state of a protein are based on thermodynamics and kinetics. One requirement is that protein folding should be a thermodynamically favorable process. The native state should have a free energy lower than the unfolded state allowing the folded protein to be stable under physiological conditions. Kinetically, a protein should be able to fold into the native conformation within a reasonable time frame. Due to the complexity of this process, an exact description of the process of protein folding has eluded scientists for a long time. Christian Anfinsen and co-workers [Anfinsen 1973] performed a series of experiments which contributed to the initial understanding of protein folding. These experiments included the folding studies of unfolded ribonuclease upon oxidation and removal of the denaturant. Two general conclusions were reached. First, that protein folding was reversible and secondly, all the information for protein folding resided in the primary sequence. However, the detailed nature of the protein folding pathway still puzzled researchers for a long time. Only recent advances in both experiment and theory have put some light into the thermodynamics and kinetics of protein folding. In the next sections the thermodynamic and kinetic requirements of protein folding are discussed in some details.

### A. Thermodynamics of protein folding: Enthalpy and Entropy

It has been determined that the free energy of the native state of a protein is only slightly lower than the unfolded state under physiological conditions [Dobson, Sali et al. 1998]. Looking at the difference in free energy, $\Delta G$, between the native and the unfolded states, defined as $\Delta H - T\Delta S$ (where $\Delta H$ and $\Delta S$ are the corresponding enthalpy and entropy change), readily provides the reason for this phenomenon. While the enthalpy change is favored for the folded state due to the various favorable interactions (e.g. the hydrophobic effect and hydrogen bonding that involve both van der Waals and electrostatic interactions), the entropic contribution to folding is generally unfavorable since it involves transitions from a large number of unstructured conformations to a single ordered structure. To fold to its native state, proteins must therefore overcome conformational entropy loss arising from the various interactions in the native state. The values of $\Delta S$ and $\Delta H$ vary from one protein to another but the result appears to be largely compensatory with the free energy change ($\Delta G$) values confined to a narrow range generally between 3.6 to 9.6 kcal/mol. With the development of supersensitive calorimetric methods, the contribution of these individual interactions to the total free energy change can be estimated.

## 1. Contribution of Hydrogen Bonding and Charge-Pair Interactions

The contribution of hydrogen bonding to the globular protein stability is widely debated [Myers and Pace 1996]. Theoretical studies by, Honig & Yang [Honig and Yang 1995] and Lazaridis [Lazaridis, Archontis et al. 1995] have suggested that polar groups contribute little or nothing to the protein stability. However, recent experiments which study the stability of various mutant proteins have suggested otherwise.

Quantum mechanical calculations suggest that the free energy of a hydrogen bond in an α-helix is 4.9 kcal/mol [Myers and Pace 1996]. This value is in agreement to the change in enthalpy observed for a neutral hydrogen bond in vapor phase. However, during protein folding in water, this value changes dramatically due to various interactions of water with the peptide bond. In the unfolded state, the polar groups of a protein interact with water via hydrogen bonding which stabilizes the unfolded state. Upon folding of a protein to its native state, these water molecules are released as the polar groups get buried inside the protein where the environment is like a non-polar solvent. Hence, the effect of the hydrogen bond was first studied by measuring the free energy of transfer of model peptide hydrogen bonds (like cyclic dipeptides or N-methylacetamide) from water into non-polar solvents. It was detected that the free energy of transfer, $\Delta G_{tr}$, of a peptide group (-NH-CH-CO-) from water to a non-polar solvent was unfavorable. This free energy of transfer varied with the non-polar solvent and its values were 5.3, 1.1 and 0.6 kcal/mol in cyclohexane, n-octanol and $CCl_4$, respectively.

Model studies also point to the fact that these forces, though important, are not the dominant forces that fold protein in water. Several examples, including solvent denaturation experiments, point to this conclusion. It is reasoned that if hydrogen bonding was the dominant force in protein folding, then solvents that form strong hydrogen bonds with the peptide backbone should completely unfold the protein, while solvents that do not affect hydrogen bonding will have no effect on protein stability at all. However, the contrary is observed in most cases. For example, methanol which is more hydrophobic than water is expected to stabilize proteins. However, it destabilizes proteins at low concentrations and this is inconsistent with the theory that hydrogen bonds are the principal driving force for protein stabilization.

Mutational studies regarding hydrogen bonding in proteins involved replacing one partner of a hydrogen bonding pair with a residue incapable of hydrogen bonding and were performed by a number of groups [Fersht and Serrano 1993; Matthews 1995]. Measurements in the differences of stability between mutant and wild type protein gave a good insight into the contribution of hydrogen bonds to protein stability. Measurements showed that the change in conformational stability, $\Delta(\Delta G)$, for the mutation of a hydrogen bonded residue to one incapable of hydrogen bonding suggested a stabilization of 2.2 kcal/mol per hydrogen bond [Myers and Pace 1996].

However, understanding the effect of hydrogen bonding upon protein stability is still at its infancy, requiring further work to get proper insights.

### 2. Contribution of Hydrophobic Interactions

Model studies, theoretical simulations and mutational studies have been performed to get an insight into the contribution of hydrophobic interactions to protein stability. It is generally agreed that most of the thermodynamic stability of proteins arises from hydrophobic interactions. The hydrophobic effect contributes approximately 1.9 kcal/mol per residue on average [Robertson and Murphy 1997]. Upon protein folding, hydrophobic side chains are buried inside the protein structure and are shielded from water. Such packing leads to favorable van der Waals interactions within the non-polar side chains, particularly among the aromatic amino acid residues. Further during folding, ordered water molecules that are specifically arranged in the unfolded state are released from this state and are free to move with the result that the entropy increases and makes favorable contributions to the $\Delta G$ term. Therefore, the contribution of hydrophobic interactions on the stability of the folded state of a protein represents the sum of the van der Waals interactions between the non-polar groups and their hydration effects. It has been proposed that the hydrophobic effects of aromatic and aliphatic groups are based on different mechanisms. In case of aromatic groups the hydrophobic effect is provided by strong van der Waals interactions between these groups and the hydration effect only decreases its value. However, in case of aliphatic groups the van der Waals interactions are enhanced by the positive hydration effects. Studies on T4 lysozyme mutants revealed that the mean "hydrophobic" $\Delta G$ of the buried surface of aromatic and aliphatic residues is approximately 28 and 40 cal/mol$Å^2$, respectively, at 25 °C [Dobson, Sali et al. 1998].

Taken together, it can be estimated that the major contribution to thermodynamic stability arises from hydrophobic interactions. For example, theoretical calculations on lysozyme reveal that the non-polar groups contribute 250 kcalmol$^{-1}$ and the polar groups contribute 87 kcal mol$^{-1}$ at 25 °C to the free energy of denaturation [Dobson, Sali et al. 1998]. The stabilization of the native state due to the effective energy term (about 537 kcal mol$^{-1}$) is counterbalanced by the configurational entropy contribution of about 523 kcal mol$^{-1}$ at 25°C. This yields a net free energy of unfolding of 14 kcal mol$^{-1}$ which is a typical value for globular proteins.

## B. Kinetics of Protein Folding

In addition to the thermodynamic contribution, a kinetic contribution is based on the demand that the denatured polypeptide chain can adopt its native conformation within a reasonable time frame. This time frame would be orders of magnitude shorter than what is required for a random search. To estimate the time required for a random search, let us consider a polypeptide of 100 amino acids and assume that there are three configurations possible for each amino acid. In such a scenario, statistically it would adopt $10^{49}$ conformations. If only $10^{-11}$s were required to convert one conformation into the other, a random search of all conformations would still require $10^{36}$ s or about $10^{29}$ years. However, it has been experimentally determined that domains with less than 100 residues fold over much faster time scales ranging from 100 ms to 1 s. This problem is commonly known as Levinthal's paradox. Levinthal's argument led to a search for folding "pathways"(Creighton 1993). The early view of the protein folding was based on simple phenomological kinetic models. These kinetic models were drawn from simple or multiple exponential time decays of optical properties that monitor changes in protein structures upon initiation of folding. More extensive work was then carried out to identify stable intermediates, known as molten globules, to shed more light onto the exact description of folding. The challenge was to elicit a single mechanism which would explain the folding of all proteins. This led to a multitude of proposals (Figure 2) over the years which are discussed below.

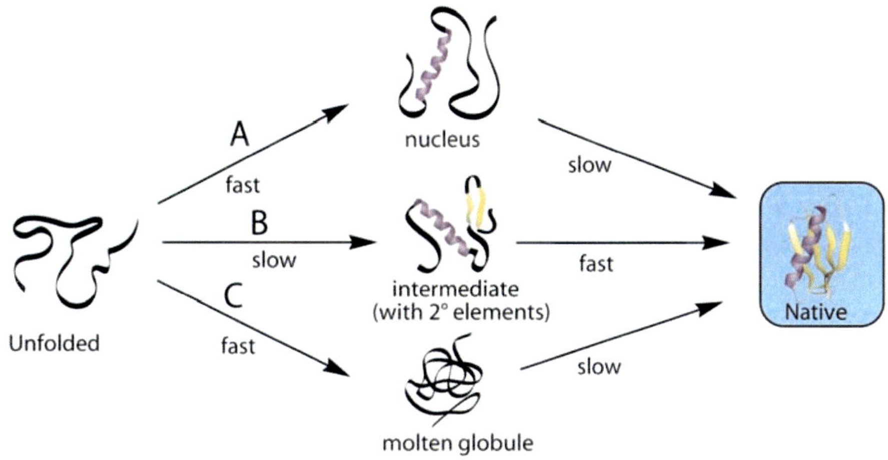


Figure 2. A schematic view of protein folding showing the three classical models: A) framework B) nucleation-growth, and C) hydrophobic collapse. The nucleation-condensation model is a combination of the framework and hydrophobic collapse model. The main differences between these models are the structure present in the intermediate and the kinetics of intermediate formation.

## 1. Nucleation-Growth Model

This model, developed in the 1960's, proposed that the tertiary structure propagates rapidly from an initial structure of local secondary structure [Wetlaufer, 1973]. However, this model fails to involve the presence of intermediates known as molten globules. During studies of protein folding it was observed that the intermediates had some of the properties of the native fold. This state became known as the molten globule and is characterized by a compact structure, containing most elements of secondary structure although the tertiary contacts are lacking.

## 2. The Framework and Hydrophobic Collapse Model

In the framework model [Kim and Baldwin 1982], the secondary structure is proposed to fold first, followed by the docking of the pre-formed secondary structural units to yield the native, folded protein. In the hydrophobic collapse model [Tanford C. 1962], the hydrophobic amino acids collapse to exclude water and form a compact structure. This collapse leads to folding in a compact volume leading to a constrained conformational search to the native state. The two alternative models to the nucleation growth model were both based on work on short alpha helical peptides. The formation of segments of helical structure occurs with time constants around 1 μs. The reactions proceed with a slow initial stage as the polypeptide chain struggles to form hydrogen bonds between a donor and an acceptor separated by three or four residues in a disordered chain. This initial stage is followed by faster rates in which the helix progressively elongates. Based thereon it was believed that proteins first form the secondary structure (like alpha helix) and then these secondary structures generate tertiary contacts which ultimately leads to the protein's folding into its native structure. Studies on predominantly α-helical proteins like Protein A corroborates this model. However there were two major drawbacks. First, formation of β-structures require more than one element for secondary structure of stability which this model fails to explain. Further, in reality, very strong conformational preferences are rare. In fact most peptides which form regular secondary structures in proteins are >90% disordered in small peptides containing the identical protein sequences [Epand and Scheraga 1968]. Consequently, hydrophobic and other interactions with the rest of the protein stabilize the unstable elements of the secondary structure.

The support of the hydrophobic collapse model came from the fact that hydrophobic driving forces provided by the expulsion of water from the burial of non-polar surfaces are substantial. However, the limitation of this model is that a non-specific hydrophobic collapse into a stable intermediate would hinder the further reorganization of both the polypeptide and side chains. Further, both models do not take into account that proteins have a 'residual structure' in the unfolded state which can be beneficial for protein folding by limiting its search space and biasing it towards the native structure.

## 3. The Nucleation-Condensation Model

Studies by Uversky and Fink [Uversky and Fink 2002], who analyzed the conformational properties of 41 native and partially folded states, found no evidence for either compact intermediates lacking secondary structure or highly ordered secondary structure in much expanded states as proposed in the models described above. This led to the proposal of the nucleation-condensation model [Fersht 1995; Daggett and Fersht 2003] by Fersht and co-workers which unites features from both the framework and hydrophobic collapse model. The original model is based in folding experiments with wild type and mutant chymotrypsin inhibitor 2 (CI2) reflecting differences in stability and activation energy [Otzen, Itzhaki et al. 1994]. Nucleation-condensation invokes the formation of long range and other native hydrophobic interactions in the transition state to stabilize the otherwise weak secondary structure. In the transition state, the secondary structural elements like the α-helices which have small configurational preferences are stabilized by the tertiary interactions made with the rest of the protein. The transition state resembles a distorted form of the native structure, with the least distorted part being called the nucleus and the distortion tending to increase with increasing distance from the nucleus.

## 4. The Energy Landscape Theory

A 'new view' of protein folding has emerged in the last few years due to advances in experiments which allow to look at atomic levels and theoretical simulations using simple statistical mechanical models [Dill and Chan 1997; Dobson, Sali et al. 1998; Brooks 2002; Cheung, Chavez et al. 2004]. This new view, known as the energy landscape theory', refutes the classical view of protein folding which describes it as following a nearly sequential series of discrete intermediates. In contrast, the energy landscape theory considers folding as the progressive organization of an ensemble of partially folded structures through which the protein passes on its way to the natively folded one. As put elegantly by Dill [Dill and Chan 1997], "In the metaphor of a energy landscape, folding is seen as more like the trickle of water down mountainside of complex shapes, and less like the flow through a gulley". The path traveled by the protein to fold can be described by a contoured energy landscape that directs the unfolded protein towards the final (low energy) conformation by a 'funnel' (Figure 3). On initiation of folding, the protein can be considered as a ball traveling through a rugged landscape following some particular trajectory while being randomly directed by Brownian motion. The shape of the landscape directs all conformations towards the native state (the bottom of the 'funnel') yet avoids the need to sample all of the possible conformations. The funneled organization of the energy landscape dominates the kinetics of folding. First, the funneling suggests that the process of folding will be hard to destroy by mutation as long as stability is maintained. Mild mutation will initially shift the folding routes and, if one set of routes to the native structure is completely blocked by a very destabilizing mutation, another set of routes will take over, as long as the final structure is stable. Second, funneling requires that a bottleneck must block an entire set of folding routes. This limits the space for the protein to search for the proper folding path. However, there must also be ensembles of structures through which the system must pass to fold. These are called 'transition state ensembles'. They consist of high-energy intermediate structures. The transition state ensembles derive their stability from interactions present in the native structure. The protein would finally fold into the bottom of the funnel through this 'transition state ensemble'.

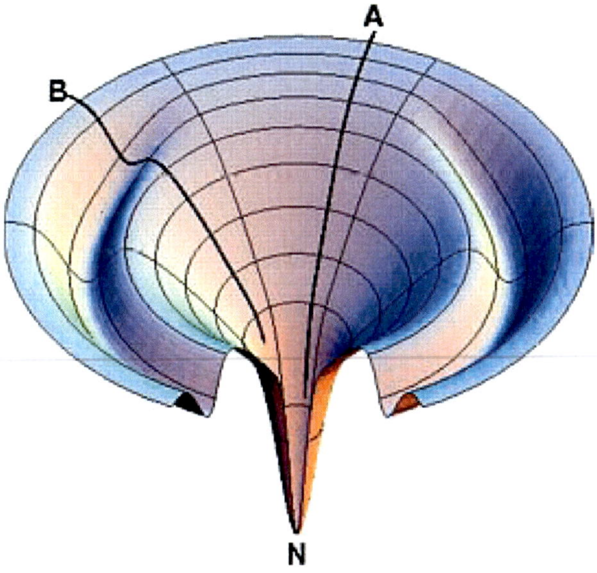


Figure 3. Image of an energy landscape that represents the reduction of conformational entropy as protein chains fold to their native state, N. A and B are two possible pathways in which pathway B shows an intermediate formation on the road to the native state N. Adapted with permission from http://www.dillgroup.ucsf.edu/.

## III. Self-assembly of Higher Order Protein Structures

The structure of a protein is the result of an ideal partitioning of nonpolar and polar side chains of individual polypeptides between regions revealing a high and low dielectric constant in the solvent environment [Jaenicke, 1987]. In aqueous solution the hydrophobic surface area of a protein is minimized as a consequence. However, many proteins are dimers, trimers, tetramers, or even higher order assemblates of identical or different polypeptide chains. The basic principle of structure formation is also valid for such oligomeric proteins. Each protein subunit is often folded into an apparently independent globular conformation (although exceptions exist), which then interacts with other monomers. The centers of the interfaces between monomers look usually similar to the interiors of individual monomers: they are closely packed and involve primarily hydrophobic interactions between non-polar side chains. The periphery of the interface, however, is in principle similar to the surface of monomers and displays hydrogen bonds and salt bridges between ionized side chains.

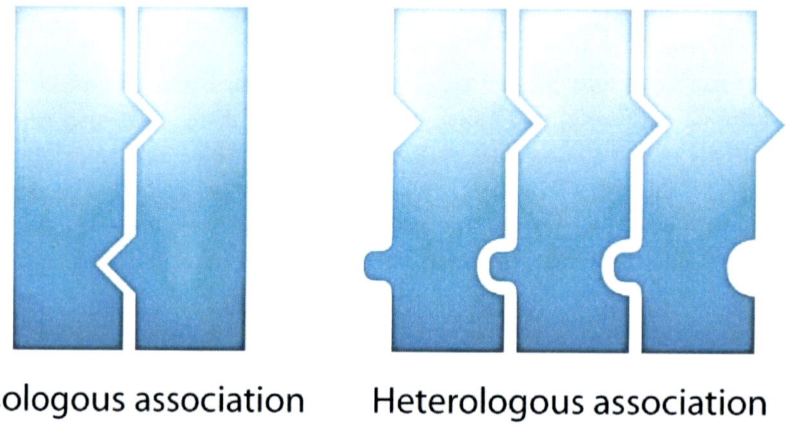


Figure 4. Schematic illustration of isologous association to form a dimer and a heterologous association in formation of a tetramer.

Two fundamental types of interaction between identical monomers are possible, which have been named isologous and heterologous (Figure 4) [Creighton, 1990]. Isologous association involves identical surfaces on both monomers, which associate to produce a dimer with a two-fold rotation axis of symmetry. The two monomers are equivalent, which requires that both monomer interfaces are complementary about a mirror plane that coincides with the symmetry axis of the dimer. Non-equivalent association of two surfaces to form a non-symmetrical dimer is unlikely because some of the complementary sites would not be paired. No further association is possible using the binding sites of isologous interactions. Association to produce tetramers requires the monomers to have an additional binding surface. Most tetramers are formed by two sets of isologous interactions to give three two-fold axes of symmetry. Heterologous association involves two different sites that are complementary and largely non-overlapping. In such a dimer, the two monomers are not equivalent and each has one unpaired binding site, which tends to bind to another monomer. Indefinite self-assembly results unless the geometry of the association leads to a closed ring (Figure 4). Any oligomeric protein with a fixed number of identical subunits that is not a power of two is likely to result from heterologous association. Some elongated structural proteins, such as actin, are formed by indefinite heterologous self-assembly of globular monomers.

## A. Actin Assembly

Actin, which is the most abundant protein in many eukaryotic cells, forms filaments with diameters of 8 nm. The filaments consist of a tight helix of uniformly oriented globular actin molecules (G-actin) and reflect polar structures with two structurally different ends. The so called minus end is relatively inert and slowly growing, while the so called plus end is faster growing [Kabsch & Vandekerckhove, 1992]. Self-assembly of actin *in vitro* requires ATP as well as both monovalent and divalent cations, such as $K^+$ and $Mg^{2+}$. Actin itself is an ATPase. Under physiological conditions, $Mg^{2+}$ATP bound G-actin is incorporated into growing filaments at the plus end. ATP-actin is then converted in ADP-actin by slow hydrolysis as actin monomers are shifted along the filament toward the minus ends. Thus, actin assembly at steady state can be described by an ATPase cycle featuring the energetic imbalance between the fast and slow ends of filaments linked to the irreversible hydrolysis of ATP. This cycle comprises three elementary steps: (a) net depolymerization from the minus end, releasing ADP-G-actin; (b) exchange of ATP for bound ADP, restoring ATP-G-actin, which (c) undergoes net assembly at the plus end. The kinetic constants of all these steps determine both the concentrations of each species at steady state and the turnover rate of actin filaments [Disanza 2005]. It is noteworthy that the concentration of ATP-G-actin, which is kept stationary during this cycle, is intermediate between the critical concentrations at the plus and minus ends. Importantly ATP hydrolysis weakens the bonds in the polymer (F actin) and thereby promotes disassembly [Kabsch & Vandekerckhove, 1992].

## B. Silk Assembly

Other intensely studied protein filaments are silks. Arthropods produce a great variety of silks, which are used in the fabrication of structures outside of the body of the animal. Silks are typically fibrous filaments or threads with diameters ranging from a few hundred nanometers to 50 μm. Insects and spiders use their silks for life lines, nests, traps and cocoons. In case of silks assembly does not start with globular folded monomers, but rather with intrinsically unfolded proteins. Characteristically, the silk is produced in specialized glands and stored in a fluid state in the lumen of the gland. Within this fluid state, the silk proteins reveal almost entirely a random coil structure. As the fluid passes through the spinning duct, a rapid transformation to the assembled state takes place and the silk becomes water-insoluble [Scheibel, 2004 a]. Importantly, in contrast to the fluid state, the final higher-order assembly state has a high content of regular secondary structure. The control of protein folding is therefore crucial for silk assembly. This requires a well-designed molecule with bi-stable folding and careful control of the pH, ionic concentration and composition, and water content.

Wide variations in composition and structure have been found among different silks. Therefore, the classification of silks is often based on the predominant form of regular secondary structure in the final filament. Silks are classified as beta silks (which are rich in β-sheets), alpha silks (which resemble the structural features of α-keratin), Glycine-rich silks (which form a polyGlycine II-like structure), and collagen-like silks (which resemble structural features as found in collagens) [Scheibel & Serpell, 2005].

## C. Collagen Assembly

Collagens are a family of highly characteristic fibrous proteins found in all multi-cellular animals. Importantly, it has to be mentioned that also non-fibrous collagens exist, which are no further discussed in this chapter. The characteristic feature of a typical fibrous collagen molecule is its long, stiff, triple-stranded helical structure, in which three collagen polypeptide chains, called α chains, are wound around each other in a rope-like super helix [Burgeson, 1988]. All collagen polypeptides have a similar primary structure, in which every third amino acid is Glycine (Gly-Xaa-Yaa), with a predominance of Proline residues at position Xaa and Yaa. Many of the Proline residues, as well as occurring Lysine residues at the position Yaa, are hydroxylated as a result of posttranslational modification. The repeating sequence of the collagen chain is reason for the three-dimensional structure of collagens, which has a slightly twisted, left-handed, threefold helical conformation like that of polyProline II and polyGlycine II and entirely different from that of an α-helix. Collagens of 1000 amino acid residues produce a triple helix 1.4 nm in diameter and 300 nm in length.

Collagen is synthesized in the precursor form, procollagen, which contains non-triple helical extensions at both ends. Removal of the amino (N) - and carboxyl terminal (C) - propeptides by specific proteases occurs prior to final fibril assembly. The C-propeptides are essential for the initiation of procollagen molecular assembly from the constituent chains [Engle and Prockop, 1991]. Removal of the N-propeptide initiates a first initiation of lateral association. Thereafter, the C-propeptide is cleaved inducing further assembly of procollagen into fibrils [Fleishmajer et al., 1983]. Collagens then assemble side-by-side into microfibrils, which are thin in diameter (10-300 nm) and hundreds of micrometers long. The collagen fibrils often assemble further into larger, cable-like bundles with several micrometers in diameter [Burgeson, 1988].

## D. Viruses

Other self-assembly interactions occur in higher-order structures, such as icosahedral viruses. An icosahedron, which has 60 monomers in identical symmetrical positions, is the largest symmetrical closed structure that is possible with identical monomers. However, the shells of many icosahedral viruses are composed of more than 60 monomers, excluding that all have equivalent positions. In particular, two-, three-, and five-fold symmetries are present between subunits. This structural quasi-equivalence is possible because the virus-coat protein molecule has several domains that can move relative to each other and thus can adopt these three packing arrangements.

## E. Hemoglobin Assembly

In addition to homomultimers, in which all of the subunits are identical, heteromultimers can assemble upon two or more subunits that differ in their primary structure. One well understood example of such heteromeric self-assembly and conformational switching process is that of hemoglobin. This oxygen-carrying tetrameric protein is composed of two separate monomers, α and β, which both occur twice in the active protein (Figure 5). Although each subunit is perfectly capable of reversibly binding oxygen, this ability is dramatically modified in the $\alpha_2\beta_2$ structure. Association of the first oxygen molecule with hemoglobin markedly enhances the affinity of hemoglobin for second oxygen, which in turn promotes the binding of a third, and so on. This is a consequence of conformational changes that occur when oxygen binds to the individual subunits. The practical significance of this behavior is that the predominant *in vivo* forms of hemoglobin are the fully oxygenated and fully deoxygenated states. Such effect provides the basis for efficient oxygen binding in the lungs and completes oxygen release to the tissues. The nature of the interactions between the individual subunits in hemoglobin can be found when one compares the structures of the α and β subunits of hemoglobin with that of myoglobin. The latter is the oxygen storage protein in muscle and is completely monomeric. Not unexpectedly, the gross structures of these proteins are very similar. However, there are some significant differences. Although there is little contact between the α and α, or β and β components of hemoglobin (Figure 5), there are extensive non-covalent interactions between the α and β subunits in the hemoglobin tetramer. The so-called "packing contacts" between the α and β chains primarily involve three α-helices. Most of these contacts are hydrophobic in nature. In marked contrast, the corresponding residues in myoglobin are decidedly polar. Although the hydrophobic residues that line the interface between the α and β chains play a critical role in promoting the association of these

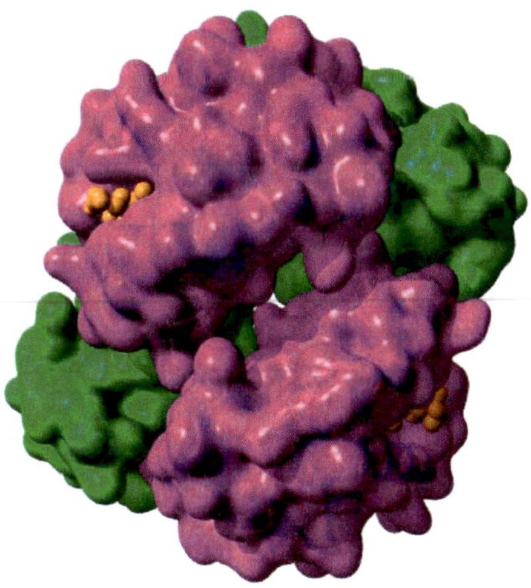

Figure 5. The crystal structure of tetrameric human deoxyhaemoglobin [Fermi *et. al* (1973). *J. Mol. Biol. 175*, 159-174] (PDB ID: 4HHB). The heme moiety is seen in orange, the α-subunits are shown in pink and the β-subunits are shown in green.

hemoglobin subunits, the corresponding polar residues in myoglobin enable this monomeric protein to readily interact with a significantly more polar environment. There are also intersubunit contacts in hemoglobin that undergo changes upon oxygen binding (which constitute the conformational changes that control oxygen affinity). These non-covalent interactions are primarily hydrogen bonds and salt bridges and are termed "sliding contacts", since they differ in the oxygenated and deoxygenated forms.

Clearly, the multitude of precisely positioned non-covalent interactions that exist between the individual subunits in hemoglobin control both structure and function. It is possible to convey a sense of how influential the non-covalent interactions are in producing a functionally active supramolecular complex by examining those situations where simple structural errors lead to lasting consequences. This is especially evident in the case of aberrant human hemoglobins. Hemoglobin in patients suffering from sickle-cell anemia contains a Valine at a position on the β chain normally reserved for Glutamic acid. This has little effect on the structure of the oxygenated form of hemoglobin. However, the corresponding deoxyhemoglobin exhibits a profound tendency to aggregate, forming long fibers that deform the cell membrane (i.e. sickling). These aggregated hemoglobin mutants are another example of self-assembly, albeit an inappropriate one. Erythrocytes containing these aggregates are not only deformed but are more rigid than their normal counterparts. As a consequence, they often get trapped in capillaries, resulting in both pain and inflammation. In addition, these sickled red blood cells are relatively fragile and must be replaced with greater frequency than healthy erythrocytes [Dickerson & Geis, 1983].

## IV. Conformational Switches of Proteins Leading to Disease

The ability of proteins to fold into a defined and functional conformation is one of the most fundamental processes in biology. Certain conditions, however, initiate misfolding or unfolding of proteins. This leads to the loss of functional protein or it can result in conformational switches followed by self-assembly and aggregation, as already described in the case of hemoglobin.

Typically, specific intermolecular interactions between hydrophobic surfaces of structural subunits in partially folded or unfolded intermediates are responsible for the formation of improper assemblies and aggregates. The aggregation propensity of a given protein correlates in part with the lifetime of partially folded intermediates and with the exposure of its hydrophobic amino acids. Longer lived intermediates are more likely to cause aggregation, since there is a greater chance of interaction with another partially folded intermediate. An important consequence is that aggregation will be favored by factors and conditions which favor the population of these intermediates, and that aggregation will be influenced by properties of the intermediates [Uversky, 2003; Jaenicke, 1995; Goldberg *et al.*, 1991]. The initial stages of aggregation are quite specific in the sense that they involve the interaction of surface elements of one molecule with matching hydrophobic surface areas of a neighboring molecule. The three-dimensional propagation of this process leads to large aggregates. Initially, the aggregates will be soluble, but eventually their size will exceed the solubility limit. Solvent-exposed hydrophobic surfaces will further minimize their solubility. Folding intermediates are more prone to aggregate than the unfolded state (although exceptions exist), because in the unfolded state the hydrophobic side chains are scattered relatively randomly in many small hydrophobic regions. In the partially folded intermediates there will be large patches of contiguous surface hydrophobicity, leading to a much stronger propensity for aggregation. In the productive folding process, these surfaces of folding intermediates will interact in an intramolecular manner to form the native conformation.

Aggregation often appears to be irreversible, but this is a reflection of the very slow rates of disaggregation and the fact that the equilibrium lies far in favor of the aggregate rather than its soluble monomeric form [Fink, 1998]. Under certain conditions aggregates can be reversed. In practice, however, once insoluble aggregates form, the process is irreversible under native-like conditions *in vitro* and *in vivo*. Circumstances that lead to the population of partially folded intermediates, especially if their concentration is high, are thus likely to lead to aggregation. In most instances of aggregation, there is a kinetic competition between aggregation and folding. While proper folding is favored at low protein concentrations, high concentrations of folding polypeptide chains will favor aggregation as a higher order process [Kiefhaber *et al,* 1991].

The aggregation potential of a protein can be influenced by genetic defects and mutations that lead to destabilization of the native state relative to the partially folded intermediate. In the context of physiological aggregation, the role of post-translational processing might also be critical. Further, adverse cellular and physiological factors or environmental conditions influence the aggregation behavior of proteins. Among the cellular conditions that have been identified to act as determinants for protein aggregation, elevated temperature is of key importance [Uversky *et al,* 2001; Lomas *et al.,* 1992]. In addition to increased temperature, altered pH [Dobson 2001] and decreased ATP levels may inhibit the acquisition of the folded state of proteins and promote the transition to partially folded intermediates, thus increasing the aggregation propensity and pathogenecity [Gregersen *et al* 2003]. Further, oxidative stress has been shown to contribute to the pathogenesis of many protein aggregation processes [Butterfield & Kanski 2001]. Additional determining factors of the extent and rate of protein aggregation are ionic strength, co-solutes (e.g. ligands that interact selectively with either the native or the non-native conformation of the protein or the aggregated form), and the presence or absence of various molecular chaperones.

## A. Conformational Diseases

Importantly, the investigation of protein aggregation has been gained an increasing role in human health. The grown importance of protein aggregation in diseases yielded in the proposal of Carrell and Lomas (1997) that a group of disorders, including Alzheimer's disease (AD), Parkinson's disease (PD), Huntington's disease (HD), transmissible spongiform encephalopathies (TSEs) or prion diseases, cystic fibrosis, and sickle cell anemia, among others, should be named conformational diseases (Table 1) [Scheibel and Buchner, 2006; Scheibel, 2004 b; Johansson et al, 2004; Aguzzi and Haass, 2003; Stojanovic et al, 2003; Ishimaru et al, 2003; Bates, 2003; Dawson and Dawson, 2003; Stirling et al, 2003; Sandilands et al, 2002; Dobson, 2001; Merlini et al, 2001; Avila, 2000; Clark and Muchowski, 2000; Damas and Saraiva, 2000; Kelly, 1998; Carrell and Lomas, 1997; Kielty and Shuttleworth, 1994; Nishio et al, 1983]. In accordance, conformational diseases are described as conditions in which a constituent protein undergoes a structural change which results in higher-order self-assembly, aggregation, and tissue deposition [Carrell and Lomas,

1997]. In the case of protein aggregation disorders, the formation of oligomers and aggregates exerts a toxic gain-of-function effect on the cell, and the cell damage / death is decisive for the clinical phenotype [Dobson, 2001]. Strikingly, the pathogenesis in these diseases is quite different, as described by the five selected examples of $\alpha$1-antotrypsin deficiency, AD, PD, HD, and prion diseases.

**Table 1. Selection of the most common diseases based on protein aggregation.**

| Disease | Protein |
|---|---|
| $\alpha_1$-Antitrypsin deficiency | $\alpha_1$-Antitrypsin |
| AH amyloidosis | Immunoglobulin heavy chain |
| Alzheimer's disease (AD) | Amyloid precursor protein (APP)<br>A$\beta$-peptide 1-42, 1-43<br>Tau |
| Amyotrophic lateral sclerosis (ALS) | Neuronal intermediate filaments (IF) |
| Cancer | p53 |
| Cataracts | Crystallins |
| Chronic inflammatory disease | Serum amyloid A protein |
| Creutzfeldt-Jakob disease (CJD); New variant CJD | Prion protein (PrP) |
| Down's syndrome | Amyloid precursor protein (APP)<br>A$\beta$-peptide 1-42, 1-43 |
| Familial amyloid neuropathy I | Transthyretin |
| Familial amyloid neuropathy II | Apolipoprotein A1 |
| Familial cardiac amyloid | Transthyretin |
| Familial visceral amyloidosis | Lysozyme |
| Fatal familial insomnia | Prion protein (PrP) |
| Frontotemporal dementia | Tau |
| Gerstmann-Sträussler-Scheinker's disease | Prion protein (PrP) |
| Haemodialysis amyloidosis | $\beta$2-Microglobulin |
| Hereditary non-neuropathic systemic amyloidosis | Lysozyme |
| Huntington's disease (HD) | Huntingtin |
| Injection-localized amyloidosis | Insulin |
| Kuru | Prion protein (PrP) |
| Parkinson's disease (PD) | $\alpha$-Synuclein |
| Primary systemic amyloidosis | Immunoglobulin light chain |
| Reactive secondary systemic amyloidosis | Serum amyloid A protein |
| Retinitis pigmentosa | Rhodopsin |
| Senile systemic amyloidosis | Transthyretin |
| Sickle cell anemia | Hemoglobin |
| Sporadic insomnia | Prion protein (PrP) |
| Transmissible spongiform encephalopathies (TSEs) | Prion protein (PrP) |
| Type II diabetes | Islet amyloid polypeptide (IAPP) |

In $\alpha$-1-antitrypsin deficiency, a mutation hinders the proper folding of the protein in the ER of liver cells. The misfolded protein tends to form oligomers which are targeted for degradation [Carrell and Lomas, 2002]. In heterozygous carriers and in homozygous patients with the lung form of the disease, the capacity of degradation components of the protein quality-control system is sufficient to cope with the accumulated protein. However, owing to a yet unexplained decrease in the degradation capacity found in 10-15 % of homozygous patients, the protein aggregates cannot be eliminated in the liver cells of such individuals and they develop cirrhosis-like liver damage and hepatocellular carcinoma [Wu et al, 1994].

The type of pathogenesis as found in HD is shared by at least nine other inherited neurological diseases where the pathogen is a string of glutamine residues, which is part of the respective protein [Sakahira et al, 2002; Taylor et al, 2002; Wanker, 2000; Perutz, 1999]. The mutation that causes HD is an unstable expansion of CAG (encoding glutamine) in the gene Huntingtin. The mutation is inherited in an autosomal dominant manner. In patients with HD, the glutamine repeat length may be more than 55, and the longer the repeat the more prone to aggregation is the fragment. The later finding is reflected in earlier disease onset for patients with long repeats compared to patients with shorter strings of glutamine [Zoghbi and Botas, 2002].

The self-aggregating cellular proteins, α-synuclein and Aβ-peptide, are the considered pathogens in a specific subset of PD and AD [Dobson, 2001]. Early forms of the diseases are inherited due to mutations in the respective genes, which further promote the self-aggregation of the proteins. Such gain-of-function effects are in contrast to ones found in late onset forms of PD and AD, where self-aggregating proteins, due to an intrinsic conformational instability of the wild-type protein, accumulate and participate in the development of degenerative disorders.

Prion diseases or TSEs are rare fatal neurodegenerative diseases of humans and animals [Hetz and Soto, 2003; Collinge, 2001; Prusiner, 1998]. Primary symptoms include progressive dementia and ataxia [Ironside and Bell, 1997]. The hallmark pathologies of TSEs are spongiform degeneration of the brain accompanied by extensive astrogliosis and accumulation of the abnormal, protease-resistant prion protein isoform in the central nervous system, which sometimes forms plaques [Budka et al, 1995]. TSEs in humans can be divided into three groups: familial, sporadic, and infectious. Human familial TSEs are all associated with different mutations in the *PrP* gene and include some forms of Creutzfeldt-Jakob disease (CJD), Gerstmann-Straussler-Scheinker's (GSS) syndrome and fatal familial insomnia (FFI) [Prusiner and Scott, 1997]. Sporadic CJD has not been associated with any known mutation and occurs worldwide with an incidence of 0.5 – 1.5 new cases per 1 million people each year [Johnson and Gibbs, 1998]. Infectious TSE diseases include Kuru, which was propagated by ritualistic cannibalism, and iatrogenic CJD, which is spread by tissue transplantation, contamination of surgical tools or inoculation with materials derived from CJD-infected tissues [Prusiner, 1998]. New variant CJD (vCJD) is a novel infectious disease, which is strongly linked to exposure to the bovine spongiform encephalopathy (BSE) agent, the most common TSE disease in cattle [Collinge, 2001; 1999; Bruce, 2000; Wille et al, 1996].

## B. Amyloid Fibrils

Several forms of deposits can be found for aggregating proteins. The protein aggregate deposits involved in conformational diseases may have various ultrastructural organizations. In most instances the misfolded protein is rich in β-sheet conformation. The highly organized oligomerization of misfolded proteins into β-sheet-rich conformations can lead to the production of so-called 'amyloids', which describe extracellular, fibrillar protein deposits associated with disease in humans [Scheibel and Buchner, 2006; Scheibel and Serpell, 2005; Horwich, 2002]. The deposits accumulate in various tissue types such as brain, vital organs (liver and spleen), or in skeletal tissue, and are defined by their tinctorial and morphological properties [Sunde and Blake, 1997]. The quantity of such aggregates can be almost undetectable in some cases or reach kilograms of deposited protein in systemic diseases [Dobson, 2004]. Each amyloid diseases, also called amyloidosis, involves the aggregation of one specific protein although a range of other components, including other proteins and carbohydrates, is also incorporated into the deposits.

Amyloid fibrils are unusually stable, unbranched, range from 6 to 12 nm in diameter, and specifically bind the dye Congo red (CR) showing an apple-green birefringence in polarized light [Teplow, 1998; Eanes and Glenner, 1968]. Amyloid fibrils produce a characteristic 'cross-β' X-ray diffraction pattern, consistent with a model in which stacked β-sheets form in parallel to the fibril axis with individual β-strands perpendicular to the fibril axis [Sunde and Blake, 1997; Geddes et al, 1968]. Remarkably, fibrils of similar appearance to those containing large proteins can be formed by peptides with just a handful of residues. One can consider that amyloid fibrils are highly organized structures (effectively one-dimensional crystals) adopted by an unfolded polypeptide chain.

Amyloidosis are slow on-set and degenerative, and can be found not only in humans but in most mammals. It is important to mention that a single protein in addition to amyloid-like deposits can also form various other aggregate morphologies which can all cause disease. The best described samples are immunoglobulins, where light and heavy chains can aggregate and form deposits with various morphologies mostly in kidneys [Merlini et al, 2001; Preudhomme et al, 1994]. The morphologies can be fibrillar, granular, crystalline, or tubular leading to different diseases depending on the morphologies. In addition to diseases with extracellular deposits, there are others, notably PD and HD, that appear to involve very similar aggregates, which are intracellular e.g. in the nucleus or the cytoplasm, and these diseases are therefore not included in the strict definition of amyloidosis [Dobson, 2001; Rochet and Lansbury, 2000].

Although the proteins involved in amyloidosis are distinct with respect to amino acid sequence and function, the fibrils formed by different amyloidogenic proteins are structured similarly. This observation is remarkable since the soluble native forms of these proteins vary very considerably. Some of the soluble 'starter' proteins are large, some small, some are largely α-helical, some largely β-sheet. Some are intact in the fibrous form others are at least partially degraded. Some are cross-linked with disulphide bonds and some are not. For many years it was generally assumed that the ability to form amyloid fibrils with characteristics as described above was limited to a relatively small number of proteins, largely those seen in disease states, and that these proteins possess specific sequence motifs encoding the amyloid core structure. Recent studies have suggested, however, that the ability of polypeptide chains to form such structures is common, and indeed can be considered a generic feature of the backbones of polypeptide chains [Dobson, 2004]. The most direct evidence for the latter statement is that fibrils can be formed by many different proteins that are not associated with disease, including such well-known proteins as myoglobin, and also by homopolymers such as polythreonine or polylysine [Dobson, 2004].

# V. Initiation of Self-assembly and Conformational Switches of Proteins

## A. Temperature as Assembly Trigger

In case of natural polypeptides, elevated temperature is most often an unwanted assembly trigger, since, typically, high temperatures perturb the native protein conformation to a sufficient degree to promote irreversible aggregation. Importantly, it is commonly observed that aggregation starts at temperatures well below the equilibrium melting temperature of the protein. This observation suggests that aggregation is not necessarily starting with fully unfolded molecules. Rather it appears that partially unfolded protein molecules are the reactive species that form aggregates. Since temperature strongly affects reaction kinetics due to exponentially increased rate constants, the thermal kinetic energy of reactants is increased. As a result, the reactant collision frequency as well as the probability of collisions revealing enough energy to overcome activation energies rises at elevated temperatures. For diffusion-controlled reactions, elevated temperatures increase the rate of diffusion of reactant species, thus speeding up the rate of reaction. Therefore, the protein aggregation rates are much higher at elevated temperatures.

## B. pH as Assembly Trigger

pH has also a strong influence on protein aggregation rates. Proteins are often stable against aggregation over narrow pH ranges. However, they may aggregate rapidly in solutions with pH outside these ranges. Solution pH determines the type (positive or negative) and total charge of the protein, thereby affecting electrostatic interactions between the polypeptide's amino acid side chains. There are two different ways in which electrostatic interactions can affect protein stability. On one hand, classic electrostatic effects are the non-specific repulsions that arise from charged groups on a protein when it is highly charged, for example, at pH far off the isoelectric point (pI) of the protein. The number of charged groups on a protein is enlarged by increasing the acidity or basicity of the solution. This higher charge repulsion within the protein concomitantly destabilizes the folded protein conformation. Thus pH-induced unfolding leads to a state of lower electrostatic free energy. On the other hand, specific charge interactions, such as salt bridges (or ion pairs), can also affect protein conformational stability. In contrast to the nonspecific electrostatic effect, where larger numbers of charges destabilize the folded state, salt bridges stabilize it. Further, charges on protein molecules also give rise to electrostatic interactions between protein molecules. When proteins are highly charged, repulsive interactions between proteins stabilize protein solutions colloidally, making assembly processes such as aggregation energetically unfavorable. When proteins possess both positively and negatively charged groups (e.g., at pH values close to the pI), anisotropic charge distribution on the protein surface could give rise to dipoles. In such cases, protein–protein interactions could be highly attractive, making assembly processes such as aggregation energetically favorable.

Therefore, pH can serve as a useful trigger for self-assembly *in vivo* and *in vitro*. Laminin, a large extracellular glycoprotein (850 kDa), is the major component of basement membranes. Laminins actually present a family of structurally homologous isoforms, which reflect a cruciform shape composed of three different polypeptide chains that are held together by coiled-coil interactions and disulfide bonds. Laminins affect multiple biological functions, including cell adhesion, proliferation, migration, differentiation, neurite outgrowth, and tumor metastasis [Freire & Coelho-Sampaio 2000]. These functions depend directly on the supramolecular organization of the basement membrane, which is ultimately determined by the interactions between single laminin molecules and between laminin and other macromolecules. Acidification induces instantaneous self-assembly of laminin at concentrations below 60 nM *in vitro*, which is the critical concentration necessary for assembly in solution at neutral pH [Yurchenco *et al.* 1985, 1992]. The self-assembly triggered by acid pH in solution, which does not involve laminin denaturation or isoelectric precipitation, exhibits (*a*) very fast kinetics, (*b*) lack of a detectable minimum concentration for self-assembly, (*c*) requirement for divalent cations, and (*d*) destabilization by high ionic strength.

Accordingly, an amino acid residue with a pH dependent side chain can be artificially incorporated into synthetic peptides to be employed as an assembly trigger for such peptides. The important design principle, that stabilization of fibrillar dispersions requires of the order of one unit of net positive or negative charge per peptide molecule, has been used to design an 11 amino acid peptide (Gln-Gln-Arg-Phe-Gln-Trp-Gln-Phe-Gln-Gln-Gln), whose self-assembly behavior is independent of pH ($1 < pH < 10$). pH control is then incorporated by appropriately positioning a Glutamic acid side chain so that the peptide-peptide free energy of interaction in the tape-like substructure is strongly influenced by direct electrostatic forces between $\gamma$-$COO^-$ in Glutamic acid. The resulting peptide (Gln-Gln-Arg-Phe-Glu-Trp-Glu-Phe-Glu-Gln-Gln) can be switched from its nematic to its isotropic fluid state by increasing the pH. Acid-base titrations of fibrillar dispersions reveal deprotonation of the $\gamma$-$COO^-$ group of Glutamic acid, which occurs over wide ranges of up to 5 pH units. The values of the energy parameters controlling self-assembly can therefore be smoothly and continuously varied by changing pH. This enables isotropic fluid-to-nematic transitions to be triggered by relatively small additions of acid or base, typically one part in $10^3$ by volume of 1 M HCl or NaOH [Aggeli *et al.* 2003].

## C. Lipids as Assembly Trigger

Lipids also can act as triggers for self-assembly of proteins. Exchangeable apolipoproteins can convert between lipid-free and lipid-associated states. The carboxyl-terminal domain of human apolipoprotein A-I (apoA-I) plays a role in both lipid binding and self-association. The lipid-triggered conformational switch in the apoA-I carboxyl terminus has parallels to the pH dependent structural transition of hemagglutinin (HA) viral fusion protein. HA undergoes a conformational transition in which a 36-residue random-coil domain is converted to an $\alpha$-helix, extending the protein by 100 Å into an elongated membrane fusion-competent conformation in what has been termed a 'spring-loaded' trigger mechanism induced by low pH [Carr & Kim 1993]. Similar to viral fusion proteins, the random coil to $\alpha$-helix transition in the carboxyl terminus of apoA-I acts as a lipid-sensitive trigger and likely contributes to the energetics of apoA-I lipidation. This enthalpic contribution to the free energy of lipid association would not be available to proteins already assuming an $\alpha$-helical conformation. The apoA-I carboxyl terminus may not only serve as a site of initial lipid contact but also as a source of energy to drive the process of lipid association. Because of the presence of random-coil segments in many exchangeable apolipoproteins and their observed lipid-induced transition to α-helix, the lipid-triggered conformational transition (and assembly) observed in apoA-I may represent a common mechanism [Oda *et al.* 2003].

## D. Metals as Assembly Trigger

Metal ions play a crucial role in life, since they are highly important triggers of protein folding and also higher order protein assembly. Since the involvement of metals in self-assembly processes is countless, a few selected examples are presented to give insights into the large variety of metal triggered processes.

Heme and iron–sulfur clusters are the iron-containing prosthetic groups that catalyze the electron transfer in various electrochemical reactions in mitochondria. They are necessary for intracellular processes as diverse as respiration, controlling tissue oxygen level, signaling, anabolism, catabolism, synthesis of protein, and defense systems against oxidative damage. Heme and iron–sulfur clusters acquire biological activity only when they are linked to a polypeptide backbone, which then adopts the correct conformation. For iron–sulfur clusters three types are common constituents of proteins. The simplest type is a single iron atom coordinated to four sulfhydryl groups of Cysteine residues. The structure of the second and the third type of iron-sulfur clusters are 2Fe-2S and 4Fe-4S [Tyler 1992]. Mitochondrial electron transport complexes (ETC) I, II, and III contain iron sulfur clusters of the types 2Fe-2S and 4Fe-4S. Various proteins in these ETC complexes depend on iron–sulfur clusters for correct folding, successful assembly into the large complex, and for their activity [Bera *et al.* 1999; Nijtmans *et al.* 1998].

Beneath iron, zinc plays a crucial role in self-assembly of proteins, most of them acquiring a specific structure called zinc finger. Zinc finger proteins are among the most abundant proteins in eukaryotic genomes. Their functions are extraordinarily diverse and include DNA recognition, RNA packaging, transcriptional activation, regulation of apoptosis, protein folding and assembly, and lipid binding. Zinc finger structures are as diverse as their functions. Metal binding studies on zinc-binding domains suggested that protein folding in zinc finger peptides is coupled to metal binding. The metal binding by zinc-finger peptides involves a transition from an unfolded apoprotein to a well-folded metalloprotein (Figure 6). Protein folding and metal binding in metalloproteins are not always coupled in this way. Instead, a continuum of folding behaviors is exhibited by different types of metalloproteins, with zinc-finger domains at one extreme. At the other extreme are proteins such as the blue copper proteins, which are completely folded even in the absence of metal [Nar *et al.* 1992]. For these proteins, the metal-binding site is essentially completely pre-organized and the metal binds with little structural change. In the middle of the continuum are proteins, such as calmodulin, for which the apoprotein has a considerable structure but in which substantial structural changes occur upon metal binding [Berg & Godwin 1997].

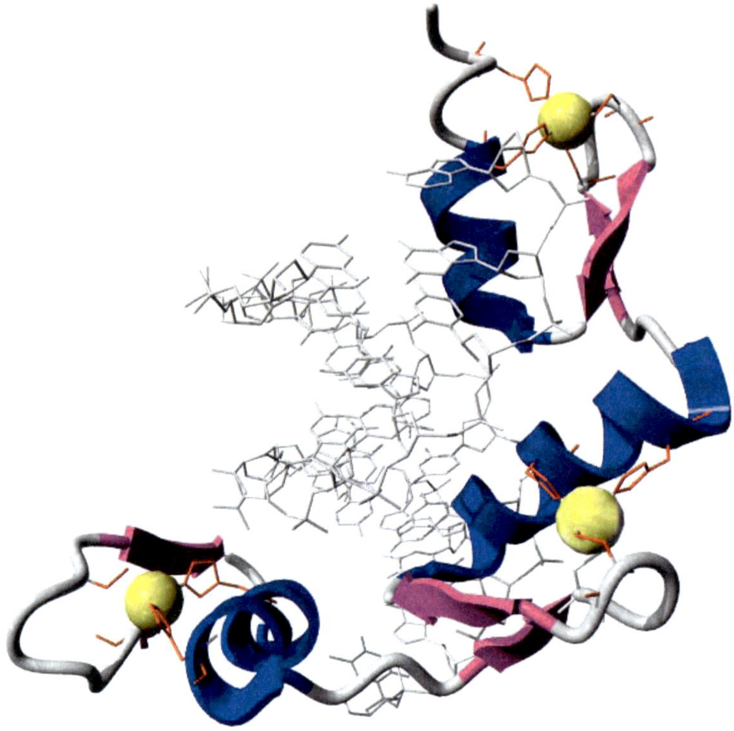

Figure 6. The crystal structure of Zinc Finger bound to DNA [Pavletich *et. al* (1991). *Science* 252, 809-817] (PDB ID: 1ZAA). The Zinc atoms are shown in yellow. Without Zinc, the peptide would not be well-folded and would not be able to specifically recognize DNA.

Strikingly, exposure to certain metal ions has been proposed as a risk factor for developing AD. The concentrations of iron and zinc in multiple brain regions vulnerable in AD are elevated in AD patients compared to their age-matched controls, whereas the copper concentrations are decreased. Imbalances of metal ions may be related to the pathogenesis of AD. Metal ions such as Al(III), Fe(III), and Zn(II) are known to promote aggregation of human Aβ *in vitro* [Kawahara *et al.* 1994]. Among them, Zn(II) induces significant Aβ aggregation at nearly physiological concentrations of the metal ion, implying participation of Zn(II) ions in the deposition of Aβ within amyloid plaques. Cu(II) is also a strong inducer of Aβ aggregation with a pH dependence different from that of Zn(II) [Atwood *et al.* 1998]. Zn(II) induces Aβ aggregation in a wide pH range (> 6.0), whereas Cu(II) is effective only at pH 6.0 - 7.0. Strikingly, several other proteins involved in neurodegenerative diseases, such as the prion protein PrP and β2-microglobulin, have also been identified to specifically bind copper and other metal ions, which also influence the assembly properties of the underlying proteins.

## E. Mechanical Force as Assembly Trigger

Mechanical forces can also be involved in reaching a highly stable assembled state of a protein. For example in spiders, successful silk assembly is based on extending, aligning and getting packed of the silk proteins in the extensional flow-field of the draw-down taper found in the distal part of the spider's spinning duct. As the hydrophobic polyalanine-rich domains, which are typical for silk proteins, align and are drawn closer together by extensional flow, they are exposed to an increasingly hydrophobic environment, which might trigger their conversion from a random coil to a β-sheet rich structure resulting in the formation of numerous interchain hydrogen bonds. Without the mechanical trigger of drawing, no ordered assembly could occur and no threads would be generated [Scheibel 2004 a].

## F. Intrinsic Sequences as Assembly Trigger

Important triggers are intrinsic sequences that often play a crucial role in protein folding, self-assembly and aggregation. Often single amino acid substitutions can cause misfolding of a protein, which can have deleterious consequences. In the case of the previously discussed conformational diseases, the formation of oligomers and aggregates exerts a toxic gain-of-function effect on the cell, and the cell damage/death is decisive for the clinical phenotype [Dobson 2001].

In contrast to sequences that trigger incorrect assembly of proteins, intrinsic sequences are also necessary to gain the correct biological function as e.g. in the case of collagen as previously described in detail. In the fibrillar procollagens, made up of an uninterrupted triple helical domain flanked by globular N- and C-terminal propeptides, folding of the procollagen C-propeptides allows interchain disulfide bonds to form. This triggers the propagation of the collagen triple helix from the carboxyl to the amino-terminus of the molecule [Hulmes 2002]. *In vitro* studies indicate that removal of the C-propeptides alone is necessary and sufficient to initiate the assembly of higher-order collagen fibrils, indicating that in this case an intrinsic sequence acts as an initiator of protein folding but at the same time as an inhibitor of higher-order self-assembly.

In biotechnological approaches, internal sequence triggers can be used to avoid premature protein self-assembly and to allow controlled assembly reactions. Recombinant spider silks have been generated from native and synthetic spider silk genes using *Escherichia coli* as the heterologous host [Scheibel 2004 a]. Upon purification, some of these recombinant silk proteins rapidly self-assemble to form insoluble microfibrils. To achieve a better regulation of the self-assembling process, sterical triggers have been incorporated into recombinant spider silks [Winkler *et al.* 1999]. Synthetic silk monomers, based on the consensus sequence from a highly repetitive dragline silk protein of *Nephila clavipes*, additionally contain Methionine residues, which can act as a redox trigger for protein assembly [Dado & Gellman 1993]. The Methionine side chain can be oxidized to the sulfoxide state in order to disrupt hydrophobic interactions between overlaying sheets, thus maintaining solubility of the protein. The procedure can be reversed by reducing Methionine sulfoxide with *N*-methylmercaptoacetamide to its original state. Another recombinant silk protein incorporates a specific phosphorylation site. Such modification can introduce charged phosphate groups into a protein. This alteration in native sequence presumably prevents hydrophobic interactions between overlaying sheets that would ordinarily lead to β-sheet formation, and thus insolubility. The protein trigger can be reactivated by enzymatic removal of the phosphate groups using calf intestinal phosphatase [Winkler *et al.* 1999]. Thus, a high level of assembly control can be achieved by incorporation of specific sequence triggers into a protein.

## VI. Nucleated and Template-Assisted Self-assembly

One of the most commonly found mechanisms of protein assembly is the phenomenon of nucleation. Protein aggregation is inherently a nucleation and growth phenomenon where aggregates accumulate, eventually exceeding their solubility and precipitate. The existence of a lag phase in the aggregation of some proteins is caused by an energy barrier towards assembly (or nucleation). The energy barrier results from the free energy required to create a new solid-liquid interface and depends on aggregate size. When the size of the new phase is above a critical size where the energy barrier is highest, growth of the nucleus occurs. Nucleation-dependent self-assembly behavior is a result of a rate-limiting nucleation step. Furthermore, the barrier to assembly may be orientationally specific. If there is an orientation with a lower free energy to assemble, then growth will occur preferentially in that orientation, resulting in ordered aggregate morphologies. For example, amyloid-like fibrils are often associated with nucleation-dependent assembly. For some protein systems undergoing nucleation dependent assembly, the rate-limiting step is the formation of a pre-nucleus species [Krishnan *et al.* 2003].

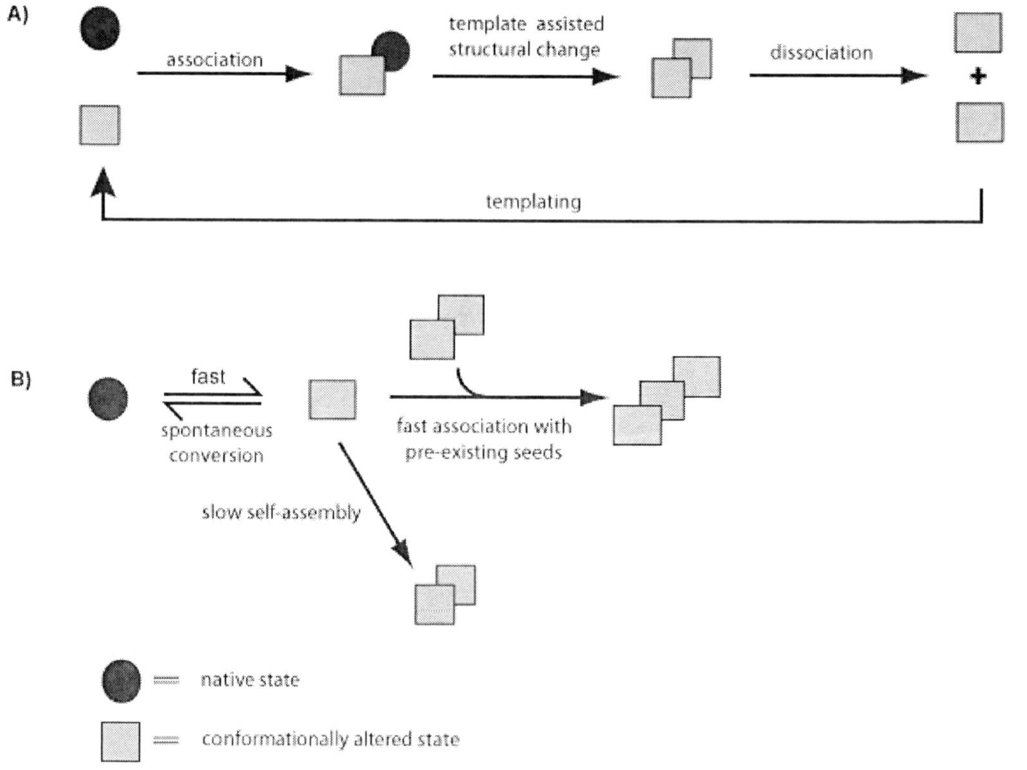


Figure 7. Schematic illustration of the template assistance model (A) and the nucleation-polymerization model (B).

In principle, two distinct mechanisms can account for such triggered nucleated self-assembly (Figure 7). In the template assistance model, the conversion from soluble into e.g. fibrillar protein is triggered by pre-existing protein fibrils acting as templates (Figure 7). In the nucleation-polymerization model, pre-existing protein fibrils serve as a seed for the abnormal polymerization of soluble but conformationally altered protein. In this model, the soluble protein can rapidly interconvert between normal and abnormal conformations and the abnormal conformer is captured and stabilized by the pre-existing protein fibril (Figure 7). Nucleation and template-assisted conversion are not mutually exclusive and a relevant mechanism could involve some combinations.

Such finding is important especially for therapeutic proteins, such as antibodies, since soluble aggregates (which can be non-covalently or covalently crosslinked) can form during long-term storage in an aqueous formulation, without any assembly into higher order aggregates or precipitates. However, if the soluble aggregate is a pre-nucleus species and it reaches a critical level (e.g. 1%), it can foster rapid assembly into large aggregates. Thus, a product that appears to have acceptable physical stability for several months could in the space of a few weeks (or even days) suddenly have unacceptable levels of aggregated protein. Similarly, if the bulk drug product already has soluble aggregates present, the long-term stability of the final formulated product may be compromised. For a given therapeutic protein product, it may become essential to have processing steps to reduce the levels of soluble aggregates (and therefore nuclei), in order to obtain acceptable long-term stability in the final formulation.

Despite the negative influences of nucleation in biological processes, it represents a perfect trigger for manufacturing new materials for certain applications. For example, recent research in the field of nanometer-scale electronics has focused on the operating principles of small-scale devices and schemes to realize useful circuits. In contrast to established "top-down" fabrication techniques, protein self-assembly is emerging as a "bottom-up" approach for fabricating nanostructured materials. Self-assembling protein fibrils provide some appropriate properties, which can be used to construct e.g. nanowire elements. Nucleated self-assembly of a prion determinant from *Saccharomyces cerevisiae* leads to chemically and physically stable 10 nm wide protein fibrils with lengths that can be controlled by the assembly conditions. Nanogold particles were covalently linked to such protein fibrils. After placing the gold-labeled protein fibers across gold electrodes, highly specific chemical enhancement of the Nanogold particles yield 50 - 100 nm wide gold wires. These bio-templated metal wires demonstrate the conductive properties of a solid metal wire, such as low resistance and ohmic behavior [Scheibel *et al.* 2003].

## VII. Conclusion

Proteins reveal the ability to fold and self-assemble rapidly to their three-dimensional structure allowing them to escape aberrant side-reactions such as unproductive folding and aggregation. In terms of energy landscapes, evolution must have shaped folding funnels to allow kinetic trapping of the native functional state. However, most proteins do not fold into a solid structure, since they have to fulfil a functional role which relies on dynamic conformational changes (e.g. upon ligand binding or membrane association). Native proteins thus are only marginally stable relative to the denatured state, and partially folded states can be formed from the folded structure by local or sub-global unfolding events.

It is not surprising that several features of protein folding and self-assembly are still not well understood. Changes in the amino acid sequence and alterations in the folding conditions influence folding as well as assembly. The most fundamental questions about the nature and frequency of different folding and assembly events, the structural properties of different ensembles, the energetic barriers between them and the shape of the multidimensional landscape, are still to be defined. Such knowledge will be a basis for elucidating the relevance of folding and assembly processes in cellular development, structure, and stability, for assessing fundamental influences leading to protein folding diseases, and for developing new biomaterials. The general concept of the multiplicity of protein folding and assembly discussed in this review may stimulate the development of new ideas and experiments to understand the underlying fundamental driving forces, leading to a deeper understanding of protein structure and dynamics.

## Acknowledgement

The authors would like to thank Lin Römer for helping to prepare the figures. The work performed in the laboratory of TS has been supported by grants from the DFG (SCHE 603/4-1; SCHE 603/4-2). SSG was supported by a fellowship of the Alexander von Humboldt Foundation.

# References

Aggeli, A.; Bell, M.; Carrick, L. M.; Fishwick, C. W. G.; Harding, R.; *et al.* (2003) pH as a Trigger of Peptide β-Sheet Self-Assembly and Reversible Switching between Nematic and Isotropic Phases. *J. Am. Chem. Soc.* **125**, 9619-9628

Aguzzi, A. & Haass, C. (2003) Games played by rogue proteins in prion disorders and Alzheimer's disease. *Science* **302**, 814-818

Anfinsen, C. B. (1973) Reasons for the folding of protein chains (Nobel lecture). *Angew. Chem. Int. Ed.* **85**(24), 1065-74.

Atwood, C. S.; Moir, R. D.; Huang, X.; Scarpa, R. C.; Bacarra, N. M. E.; et. al. (1998) Dramatic aggregation of Alzheimer A β by Cu(II) is induced by conditions representing physiological acidosis. *J. Biol. Chem.* **273**(21), 12817-12826.

Avila, J. (2000) Tau aggregation into fibrillar polymers: taupathies. *FEBS Lett* **476**, 89-92

Bates, G. (2003) Huntingtin aggregation and toxicity in Huntington's disease. *Lancet* **361**, 1642–1644

Bera, A. K.; Grinberg, A.; Bernhardt, R. (1999) A step toward understanding the folding mechanism of bovine adrenodoxin. *Arch. Biochem. Biophys.* **361**, 315-322

Berg, J. M & Godwin, H.A. (1997) Lessons from zinc-binding peptides. *Annu. Rev. Biophys. Biomol. Struct.* **26**, 357-371

Branden, C. & Tooze, J. Introduction to Protein Structure. New York: Garland Publishing, Inc.; 1991

Brooks, C. L. III (2002). Protein and peptide folding explored with molecular simulations. *Acc. Chem. Res.* **35**(6), 447-454.

Bruce M. E. (2000) 'New variant' Creutzfeldt-Jakob disease and bovine spongiform encephalopathy. *Nat. Med.* **6**, 258–259

Budka, H.; Aguzzi, A.; Brown; P.; et. al. (1995) Neuropathological diagnostic criteria for Creutzfeldt-Jakob disease (CJD) and other human spongiform encephalopathies (prion diseases). *Brain. Pathol.* **5**, 459–466

Butterfield, D. A. & Kanski, J. (2001) Brain protein oxidation in age-related neurodegenerative disorders that are associated with aggregated proteins. *Mech. Aging. Dev.* **122**, 945-962

Burgeson, R. E. (1988) New collagens, new concepts. *Annu. Rev. Cell Biol.* **4**, 551-577

Carr, C. M. & Kim, P. S. (1993) A spring-loaded mechanism for the conformational change of influenza hemagglutinin. *Cell* **73**, 823-832

Carrell, R. W. & Lomas, D. A. (1997) Conformational disease. *Lancet* **350**, 134-138

Carrell, R. W. & Lomas, D. A. (2002) Alpha1-antitrypsin deficiency: a model for conformational diseases. *N. Engl. J. Med.* **346**, 45-53

Cheung, M. S.; Chavez, L. L.; et al. (2004). "The energy landscape for protein folding and possible connections to function." *Polymer* **45**(2), 547-555.

Clark, J. I. & Muchowski, P. J. (2000) Small heat-shock proteins and their potential role in human disease. *Curr. Opin. Struct. Biol.* **10**, 52-59

Collinge, J. (1999) Variant Creutzfeldt-Jakob disease. *Lancet* **354**, 317–323

Collinge, J. (2001) Prion diseases of humans and animals: their causes and molecular basis. *Annu. Rev. Neurosci.* **24**, 519–550

Creighton, T.E. (1990) Protein folding. *Biochem. J.* **270**, 1-16

Creighton, T. E. Proteins: Structures and Molecular Properties. Edition: 2nd. New York: W. H. Freeman and Company; 1993.

Dado, G. P. & Gellman, S. H. (1993) Redox control of secondary structure in a designed peptide. *J. Am. Chem. Soc.* **115**, 12609-12610

Daggett, V. & Fersht, A. R. (2003). Is there a unifying mechanism for protein folding? *Trends Biochem. Sci* **28**(1), 18-25.

Damas, A. M.& Saraiva, M. J. (2000) TTR amyloidosis-structural features leading to protein aggregation and their implications on therapeutic strategies. *J. Struct. Biol.* **130**, 290-299

Dawson, T. M. & Dawson V. L. (2003) Rare genetic mutations shed light on the pathogenesis of Parkinson disease. *J. Clin. Invest.* **111**,145–151

Dickerson R.E. & Geis I. *Hemoglobin: Structure, Function, Evolution, and Pathology*, Benjamin Cumming Publishing Co. Menlo Park, CA, 1983

Dill, K. A. & Chan, H. S. (1997) From Levinthal to pathways to funnels. *Nat. Struct.Biol.* **4**(1), 10-19.

Disanza, A.; Steffen, A.; Hertzog, M.; Frittoli, E.; et. al. (2005) Actin polymerization machinery: the finish line of signaling networks, the starting point of cellular movement. Cell. Mol. Life Sci. **62**, 955-970

Dobson, C. M.; Sali, A. et. al. (1998) Protein folding: a perspective from theory and experiment. *Angew. Chem. Int. Ed.* **37**(7), 869-893.

Dobson, C. M. (2001) The structural basis of protein folding and its links with human disease. *Philos. Trans. R. Soc. London, Ser. B* **356**,133-145

Dobson, C. M. (2004) Principles of protein folding, misfolding and aggregation. *Semin. Cell. Dev. Biol.* **15**, 3–16

Eanes, E. D. & Glenner, G. G. (1968) X-ray diffraction studies on amyloid filaments. *J. Histochem. Cytochem.* **16**, 673-677

Engle, J. & Prockop, D.J. (1991) The zipper-like folding of collagen triple-helices and the effects of mutations that disrupt the zipper. *Ann. Rev. Biophys. Chem.* **20**,137–152

Epand, R. M. & Scheraga, H. A. (1968). The influence of long-range interactions on the structure of myoglobin. *Biochemistry* 7(8), 2864-72.

Fink, A. L. (1998) Protein aggregation: folding aggregates, inclusion bodies and amyloid. *Fold. Des.* **3,** R9-23.

Fleischmajer, R.; Olsen, B.R.; et. al. (1983) Collagen fibril formation during embryogenesis. *Proc. Natl. Acad. Sci. USA.* **80**, 3354–3358.

Fersht, A. R. & Serrano, L. (1993) Principles of protein stability derived from protein engineering experiments. *Curr. Opin. Struct. Biol.* **3**(1), 75-83.

Fersht, A. R. (1995) Optimization of rates of protein folding: the nucleation-condensation mechanism and its implications. *Proc. Natl. Acad. Sci. U.S.A.* **92**, 10869–10873.

Freire, E. & Coelho-Sampaio, T. (2000) Self-assembly of laminin induced by acidic pH. *J. Biol. Chem.* **275**, 817-822

Geddes A. J.; Parker K. D.; Atkins E. D.; Beighton E. (1968) "Cross-beta" conformation in proteins. *J. Mol. Biol.* **323**, 343-358

Goldberg, M. E.; Rudolph R.; Jaenicke R. (1991) Self-assembly of laminin induced by acidic pH. *Biochemistry* **30**, 2790-2797

Gregersen, N.; Bolund, L. et. al. (2003) Protein misfolding, aggregation, and degradation in disease. *Methods. Mol. Biol.* **232**, 3-16

Hetz, C. & Soto, C. (2003) Protein misfolding and disease: the case of prion disorders. *Cell. Mol. Life. Sci.* **60**, 133-143

Horwich, A. (2002) Protein aggregation in disease: a role for folding intermediates forming specific multimeric interactions. *J. Clin. Invest.* **110**, 1221-32

Hulmes, D. J. (2002) Building collagen molecules, fibrils, and suprafibrillar structures. J. *Struct. Biol.* **137**(1-2), 2-10.

Ironside, J. W. & Bell, J. E. (1997) Pathology of prion diseases In: Collinge J, Palmer MS (eds), Prion Diseases. Oxford: Oxford University Press, 57–88

Ishimaru, D., Andrade L,. R., Teixeira, L. S. et. al. (2003) Fibrillar aggregates of the tumour suppressor p53 core domain. Biochemistry **42**, 9022-9027

Jaenicke, R. (1987) Folding and association of proteins. *Prog. Biophys. Mol. Biol.* **49**,117-237

Jaenicke, R. (1995) Folding and association versus misfolding and aggregation of proteins. *Philos. Trans. R. Soc. London, Ser. B* **348**, 97-105

Johnson, R. T. & Gibbs, C. J. (1998) Creutzfeldt-Jakob disease and related transmissible spongiform encephalopathies. *N. Engl. J. Med.* **339**, 1994–2004

Johansson, J.; Weaver, T. E. et. al. (2004) Proteolytic generation and aggregation of peptides from transmembrane regions: lung surfactant protein C and amyloid beta-peptide. *Cell. Mol. Life. Sci.* ***61***, 326-335

Kabsch, W. & Vandekerckhove, J. (1992) Structure and Function of Actin. *Annu. Rev. Biophys. Biomol. Struct.* **21**, 49-76

Kawahara, M.; Muramoto, K.; Kobayashi, K.; Mori, H.; Kuroda, Y. (1994) Aluminum promotes the aggregation of Alzheimer's amyloid beta-protein in vitro. *Biochem. Biophys. Res. Commun.* **198**, 531-535

Kelly, J. (1998) Alternative conformation of amyloidogenic proteins and their multi-step assembly pathways. *Curr. Opin. Struct. Biol. 8*,101–106

Kiefhaber, T.; Rudolph, R.; Kohler, H. H.; Buchner, J. (1991) Protein aggregation in vitro and in vivo: a quantitative model of the kinetic competition between folding and aggregation. *Nat. Biotech.* **9,** 825-829

Kielty, C. M. & Shuttleworth, C. A. (1994) Abnormal fibrillin assembly by dermal fibroblasts from two patients with Marfan syndrome. *J. Cell. Biol.* **124**, 997-1004

Kim, P. S. & Baldwin, R. L. (1982) Specific intermediates in the folding reactions of small proteins and the mechanism of protein folding. *Annu. Rev. Biochem.* **51**, 459–489.

Krishnan S., Carpenter JF. *et al.* (2003) Oxidative dimer formation is the critical rate-limiting step for Parkinson's disease alpha-synuclein fibrillogenesis. *Biochemistry* **42**, 829-837

Lazaridis, T.; Archontis, G. et.; al. (1995). Enthalpic contribution to protein stability: insights from atom-based calculations and statistical mechanics. *Adv. Protein Chem.* **47**, 231-306.

Lomas, D. A.; Evans, D. L.; et. al. (1992) The mechanism of Z alpha 1-antitrypsin accumulation in the liver. *Nature* **357**, 605-607

Matthews, B. W. (1995). Studies on protein stability with T4 lysozyme. *Adv. Protein Chem.* **46**, 249-78.

Merlini, G.; Bellotti, V.; et. al. (2001) Protein aggregation. *Clin. Chem. Lab. Med. 39*, 1065-1075

Myers, J. K. & Pace, C. N. (1996). Hydrogen bonding stabilizes globular proteins. *Biophys. J.* **71**(4), 2033-2039.

Nar, H.; Messerschmidt, A.; Huber, R.; van de Kamp, M.; Canters, G. W. (1992) Crystal structure of Pseudomonas aeruginosa apo-azurin at 1.85 A resolution. *FEBS Lett.* **2**, 119-124

Nijtmans, L. G.; Taanman, J. W.; Muijsers, A. O.; Speijer, D.; Van den Bogert, C. (1998) Assembly of cytochrome-c oxidase in cultured human cells. *Eur. J. Biochem.* **254**, 389-394

Nishio, I.; Tanaka, T.; et. al. (1983) Hemoglobin aggregation in single red blood cells of sickle cell anemia. *Science* **220**, 1173-1175

Oda, M. N.; Forte, T. M.; Ryan, R. O.; Voss, J. C. (2003) The C-terminal domain of apolipoprotein A-I contains a lipid-sensitive conformational trigger. *Nat. Struct. Biol.* **10**, 455-460

Otzen, D. E.; Itzhaki, L. S.; et. al. (1994). Structure of the transition state for the folding/unfolding of the barley chymotrypsin inhibitor 2 and its implications for mechanisms of protein folding. *Proc. Natl. Acad. Sci. U.S.A.* **91**(22), 10422-5.

Pauling, L.; Corey, R. B. (1951) Atomic co-ordinates and structure factors for two helical configurations of polypeptide chains. *Proc. Natl. Acad. Sci. USA* **37**, 235-40.

Perutz, M. F. (1999) Glutamine repeats and neurodegenerative diseases: molecular aspects. *Trends. Biochem. Sc.i* **24**, 58–63

Preudhomme, C.; Vachee, A.; et. al. (1994) Immunoglobulin and T-cell receptor delta gene rearrangements are rarely found in myelodysplastic syndromes in chronic phase. *Leuk. Res. 18*, 365-371

Prusiner, S. B. (1998) Prions. *Proc. Natl. Acad. Sci. USA* **9,** 13363-13383

Prusiner S. B. & Scott M. R. (1997) Genetics of prions. *Annu. Rev. Genet.* **31**,139-175

Robertson, A. D. & Murphy, K. P. (1997). Protein Structure and the Energetics of Protein Stability. *Chem. Rev. (Washington, D. C.)* **97**(5), 1251-1267.

Rochet, J. C. & Lansbury, P. T. (2000) Amyloid fibrillogenesis: themes and variations. *Curr. Opin. Struct. Biol.* **10**, 60-68

Sakahira, H., Breuer, P. et. al. (2002) Molecular chaperones as modulators of polyglutamine protein aggregation and toxicity. *Proc. Natl. Acad. Sci. USA* 99 *Suppl* **4,** 16412-16418

Sandilands, A.; Hutcheson, A. M.; et. al. (2002) Altered aggregation properties of mutant gamma-crystallins cause inherited cataract. *EMBO J.* **21**, 6005-6014

Scheibel, T.; Parthasarathy, R.; Sawicki, G.; et. al. (2003) Conducting nanowires built by controlled self-assembly of amyloid fibers and selective metal deposition. *Proc. Natl. Acad. Sci. USA* **100**, 4527-4532

Scheibel, T. (2004 a) Spider silks: recombinant synthesis, assembly, spinning, and engineering of synthetic proteins. *Microbial Cell Fact.* **3,** 14

Scheibel, T. (2004 b) Amyloid formation of a yeast prion determinant. *J. Mol. Neurosci.* **23,** 13-22

Scheibel, T. & Serpell, L. (2005) Physical methods for studies of fibre formation and structure. In: Buchner J, Kiefhaber T (eds) Protein folding handbook. Vol. 3. Weinheim: Wiley, 197-253

Scheibel, T. & Buchner, J. (2006) Protein aggregation as a cause for disease. *Handb. Exp. Pharmacol.* **17**,199-219

Stirling, P. C.; Lundin, V. F.; et. al. (2003) Getting a grip on non-native proteins. *EMBO Rep.* **4**, 565-570

Stojanovic, A.; Hwang, I.; et. al. (2003) Retinitis pigmentosa rhodopsin mutations L125R and A164V perturb critical interhelical interactions: new insights through compensatory mutations and crystal structure analysis. *J. Biol. Chem.* **278**, 39020-39028

Sunde, M. & Blake, C. (1997) The structure of amyloid fibrils by electron microscopy and X-ray diffraction. *Adv. Protein Chem.* **50**, 123-159

Tanford, C. (1962) Contribution of hydrophobic interactions to the stability of globular confirmation of proteins. *J. Am. Chem. Soc.* **84**, 4240.

Taylor, J. P.; Hardy; J.; et. al. (2002) Toxic proteins in neurodegenerative disease. *Science* **296**, 1991-1995

Teplow, D. B. (1998) Structural and kinetic features of amyloid beta-protein fibrillogenesis. *Amyloid* **5,** 121-142

Tyler D.D. *The Mitochondrion in Health and Disease*, VHC Publishers, New York, 1992

Uversky, V. N. (2003) Protein folding revisited. A polypeptide chain at the folding-misfolding-non folding cross-roads: which way to go? *Cell. Mol. Life. Sci.* **60**,1852-1871

Uversky, V. N.; Lee, H. J.; et. al. (2001) Stabilisation of partially folded conformation during alpha synuclein oligomerization in both purified and cytosolic preparations. *J. Biol. Chem.* **276**, 43495-43498

Uversky, V. N. & Fink, A. L. (2002). The chicken-egg scenario of protein folding revisited. *FEBS Lett.* **515**(1-3), 79-83.

Wanker, E. E. (2000) Protein aggregation and pathogenesis of Huntington's disease: mechanisms and correlations. *Biol. Chem.* **381**, 937-942

Wetlaufer, D. B. (1973) Nucleation, rapid folding, and globular intrachain regions in proteins. *Proc. Natl. Acad. Sci. U.S.A.* **70**, 697–701.

Wille, H.; Zhang, G. F.; et. al. (1996) Separation of scrapie prion infectivity from PrP amyloid polymers. *J. Mol. Biol.* **259**, 608-621

Winkler, S.; Szela, S.; Avtges, P.; et. al. (1999) Designing recombinant spider silk proteins to control assembly. *Int. J. Biol. Macromol.* **24**, 265-270

Whitford, D. Proteins: Structure and Function, John Wiley & Sons Ltd, Hoboken, NJ, 2005

Wu, Y.; Whitman I.; et. al. (1994) A lag in intracellular degradation of mutant alpha 1-antitrypsin correlates with the liver disease phenotype in homozygous PiZZ alpha 1-antitrypsin deficiency. *Proc. Natl. Acad. Sci. USA* **91**, 9014-9018

Yurchenco, P. D.; Tsilibary, E. C.; Charonis, A. S.; Furthmayr, H. (1985) Laminin polymerization in vitro. Evidence for a two-step assembly with domain specificity. *J. Biol. Chem.* **260**, 7636-7644

Yurchenco, P. D.; Cheng, Y. S.; Colognato, H. (1992) Laminin forms an independent network in basement membranes.*J. Cell Biol.* **117**, 1119-1133

Zoghbi H. Y. & Botas J. (2002) Mouse and fly models of neurodegeneration. *Trends. Genet.* **18**, 463-471

In: Protein Folding-Misfolding…
Editors: J.P. Zbilut and T. Scheibel, pp. 35-53

ISBN: 1-60021-417-7


*Chapter 2*

# PROTEIN-PROTEIN INTERACTIONS: STRUCTURAL PRINCIPLES, STABILITY AND KINETICS OF ASSOCIATION

***Hauke Lilie****

Institut für Biotechnologie, Universität Halle, Kurt Mothes Strasse 3, D-06120 Halle, Germany

## Abstract

Most soluble proteins are biologically active in protein complexes and oligomeric structures. In order to understand the function and stability of these complexes the molecular mechanisms of protein association and the structural features of the subunit interfaces need to be elucidated. Whereas statistical analyses of these interfaces reveal the importance of hydrophobic interactions within protein contact sites, a detailed analysis of several interfaces shows diverse strategies in accomplishing specific protein-protein interactions. The hydrophobic surface of the interface might be formed as a large hydrophobic core surrounded by polar groups or as the sum of small patches interspersed by polar residues and water molecules. In higher oilgomeric complexes one might find more hydrophobic and more hydrophilic contact sites at the different levels of association. In addition to the discussion of general aspects of protein-protein interactions, protein oligomerisation based on specific structural characteristics of a given protein such as domain swapping, coiled coil structures and four-helix bundles are discussed.

## I. Introduction

The different genome sequencing projects have provided us with a vast amount of data regarding biological macromolecules (1,2). The information on the function of biological systems, however, is rather limited because the functional unit in a cell is often a complex but

* E-mail address: Hauke.Lilie@biochemtech.uni-halle.de

not a single macromolecule (RBR 3). The most detailed knowledge on protein-protein interactions is obtained from proteome studies especially in Saccharomyces cerevisiae. The functional protein interactions and dimer formation is estimated in a range of 30 000. With a proteome of ca. 6 200 proteins this indicates that every protein has several interaction partners in a cell (RBR 30,31).

In many cases the biological activity of proteins depends on their association state. The association can result in homo- or hetero-oligomers which might be either stable or only transiently formed during their specific functional cycle. The Protein Data bank (pdb) of protein structures contains more than 2000 oligomeric structures, the largest of which comprise virus shells, and homomeric or heteromeric protein complexes containing more than 20 subunits such as ferritin (24 identical subunits, (2)), the chaperonin complex GroEL/ES (21 subunits $\alpha 14\beta 7$, (3)), and dihydrolipoyl acetyltransferase (60 subunits, (4)).

All information required for the formation of the native three-dimensional structure of a given protein or protein complex is encoded in the amino acid sequence. The structure of a protein is the result of an optimum partitioning of the non-polar and polar parts of the polypeptide chain between regions of high and low dielectric constant in the solvent environment (1). In aqueous solution this leads to a minimization of hydrophobic surface area. This basic principle governs both structure formation of monomeric proteins as well as association of subunits of oligomeric proteins. This chapter deals with the structural principles of protein complexes and thermodynamic and kinetic properties of protein association.

# II. Association of Oligomeric Proteins

## 2.1. Selection Pressure for Oligomers

Oligomeric proteins are the rule, rather than the exception. Benefits of the oligomerisation may be found on the functional, structural, and genetic level. Oligomerisation permits regulation of enzyme activity. The database of Macromolecular Movements (http://molmovb.org) (10,11) depicts more than 250 distinct molecular motions of proteins, most of them based on the interaction of proteins with protein and other kind of ligands. In case of enzymes this interaction can shift the dynamic population of different protein conformations, thus regulating enzyme activity (12). In the most simple way, this leads to a switch from the tense (T) to the relaxed (R) state in the classical allosteric regulation.

These functional advantages, however, can be conferred not only by oligomeric complexes but also by large multi-domain proteins. And indeed, a few such very large multi-domain proteins consisting of more than 10 000 amino acids are known (13). Many non-ribosomal peptides are synthesized by very large proteins, e.g. the cyclosporin synthetase, in which all the different enzyme functions are located in one polypeptide chain (14). However, these very large multi-domain proteins are rare, most large protein structures are formed by association of subunits.

Oligomerisation of small gene products is by far more economical; this is quite obvious in case of viruses, where a small protein, and thus a small gene, is sufficient for building up a large virus capsid of the size of several megadaltons. Furthermore, errors occurring during protein synthesis can be eliminated easier in case of oligomeric structures. Another possible advantage of oligomeric protein complexes versus large multi-domain proteins may result from differences in folding. It has been shown that domains in the context of a larger polypeptide chain fold more slowly compared to the same domain in its isolated state (15-17). In case of the central domain of bacteriophage P22 tailspike protein, the folding of this -helix domain is accelerated about twofold upon removal of the N-terminal, structurally unrelated domain. This slightly accelerated folding increases the yield of folding significantly both in vitro and in vivo (17,18).

## 2.2. Morphology of Protein-Protein Interfaces

The growing structural data base of oligomeric proteins and protein complexes permits statistical analyses of the properties of interfaces of subunits and domains (19-22). The majority of the subunit interfaces has a surface of 1000 – 3000 $A^2$. A simple statistical survey reveals that the stability of protein association is mediated mainly by hydrophobic interactions; polar interactions, including hydrogen bonding and ionic interactions as well as surface complementarity serve as specificity factors of interaction (23,24). Recent studies suggest that, in a given interface, a few amino acids may play a dominant role in determining the free energy of stabilization of the interaction. These residues are called hot spots of the binding surface (25). In contrast to the interfaces responsible for stable subunit association, the contact sites of proteins that interact only transiently with other proteins are less hydrophobic (22).

A global statistical analysis of protein-protein interfaces, however, does not reveal information of how individual proteins use the structural principles of protein-protein interactions in oligomer formation. In order to analyse these principals in more detail Larsen and coworkers undertook a qualitative analysis of 136 subunit interfaces of homodimeric proteins on an individual basis (26). The interfaces proved to be very diverse. As a result they classified protein-protein interfaces into three groups: (i) interfaces with a well defined hydrophobic core, (ii) interfaces with a mixed character, where polar and apolar interactions as well as water molecules are evenly distributed all over the interface, and (iii) interdigitated interfaces with the complex resembling a single monomeric protein.

Out of 136 interfaces analysed 43 showed a well defined water-excluding hydrophobic core. This core is reminiscent of those in globular monomeric proteins. It is surrounded by polar interactions and structurally important water molecules (Fig. 1A).

The majority of homodimeric interfaces does not possess a well defined hydrophobic core. Instead many small hydrophobic patches are interspersed with intersubunit hydrogen bonds and water molecules (Fig.1B). These small hydrophobic patches consist of 1-6 hydrophobic residues; the overall hydrophobic surface area of these interfaces, however, might be similar to the hydrophobic core of respective subunit interfaces of similar size. Whether such an interspersed hydrophobic surface contributes to the stability of interaction in the same way as a well-defined hydrophobic core remains to be determined. Water seems to play in important role in these mixed interfaces. Hubbard and Argos (27) found that water-sized cavities cover approximately 10% of a typical interface, over half of which actually contain crystallographically observed water molecules.

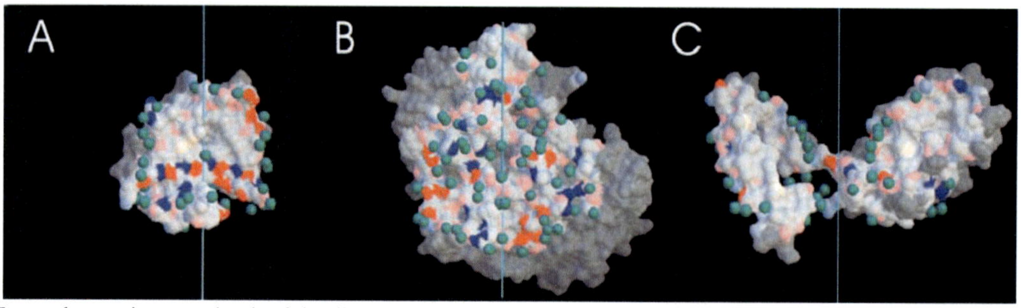


For each protein one subunit of the dimer is shown with a vertically positioned dimer axis. Carbon atoms are highlighted in white, sulfur in yellow, oxygen in red, and nitrogen in blue. Bright colors indicate hydrogen-bonded, pale colors non-bonded atoms. Water molecules are represented by green spheres. (A) Protein kinase C interacting protein 1 (PKCI-1) (pdb: 1kpa). The interface shows a large hydrophobic patch, surrounded by water molecules. At the bottom the boundary is formed by a β-sheet. (B) Adenylosuccinate synthetase (pdb: 1ade). In the interface small hydrophobic patches, polar interactins, and water molecules are evenly distributed. (C) Interleukin 10 (pdb: 1ilk). The two monomers of IL10 are highly intertwined. The interface covers a large fraction of the total subunit surface and is as hydrophobic as the interior of a protein (adapted from (26).

Figure 1. Morphology of subunit interfaces

In the survey of Larsen et al. ten out of 136 proteins analysed exhibited highly interdigitated subunit interfaces (FIg. 1C). In these examples, the two individual subunits do not form two individual globular domains, but instead, both chains together form a single globular domain. Consequently, the interfaces of these intertwined dimers resemble the interior of folded proteins. They are highly hydrophobic with packing densities of the side chains comparable to those in the hydrophobic core of monomeric proteins. In this class of proteins the individual subunits do not form stable structures by themselves. A growing number of protein-protein interactions are discovered, in which one of the partners is unstructured in its isolated state. These natively unfolded proteins can be grouped into two classes, one with random coil structures and the other with a premolten globule structure (28). Only upon association with their targets, these proteins develop an ordered and biologically active structure. Examples are several ribosomal proteins which transform to a rigid well-folded conformation only in the presence of the ribosomal RNA (29, 30). In the case of cyclin-dependent kinase inhibitor p21, the N-terminal fragment lacks stable secondary and tertiary structure in the free solution state. However, when bound to its target, Cdk 2, it adopts a well-ordered conformation (31). For the protein interaction module L27 it has been demonstrated that the structurally well-defined heterodimer of the interaction domains LIN-2 and LIN-7 assembles with an affinity in the low micromolar range whereas both domains are unstructured in their isolated state (32).

Going from dimers to oligomers it is evident that those higher oligomers possess topological units, e.g. the dodecameric aspartate transcarbamylase (ATCase) from *E.coli* consists of two catalytic trimers and three regulatory dimers. Tetramers of glyceraldehyde-3-phosphate dehydrogenase and yeast pyruvate decarboxylase are formed as dimers of identical dimers. This topology, generally observed for homotetramers, is a consequence of the association process of the respective oligomers where first two monomers form a dimer and subsequently two dimers assemble to the native tetramer. Compared to other tetramerization pathways such as the addition of monomers to dimers or trimers, the specific topology of a tetramer as a dimer of dimers leads to kinetics of association that seem to be less prone to competition with off-pathway reactions of non-native assemblies or accumulation of association intermediates (33). In this context the subunit interfaces of the monomers and the dimers might be quite different. This is illustrated by the tetrameric yeast pyruvate decarboxylase (PDC). The topology of PDC is reminiscent of a dimer of dimers (Fig. 2A). The interface between the dimers, that are forming the tetramer, is largely polar (Fig. 2B) and the tetramer spontaneously dissociates to dimers upon dilution. The dissociation constant of this dimer-tetramer equilibrium is 8.1 μM (34). The interface between the monomers, however, is strongly hydrophobic (Fig. 2B); the dimer does not dissociate into monomers just by dilution of the protein. Only in the presence of low concentrations of denaturant folded monomers can be populated (34). Unfortunately, there are only very few cases where thermodynamic, kinetic, and crystallographic data of a certain protein can be combined in order to assess a detailed structure-function relationship of higher oligomeric proteins.

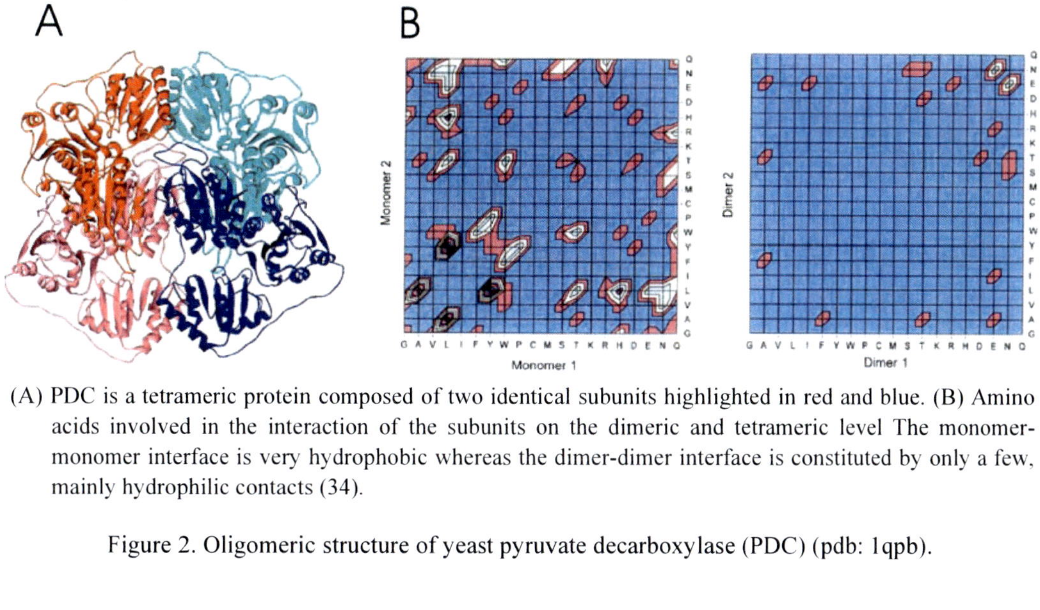


(A) PDC is a tetrameric protein composed of two identical subunits highlighted in red and blue. (B) Amino acids involved in the interaction of the subunits on the dimeric and tetrameric level The monomer-monomer interface is very hydrophobic whereas the dimer-dimer interface is constituted by only a few, mainly hydrophilic contacts (34).

Figure 2. Oligomeric structure of yeast pyruvate decarboxylase (PDC) (pdb: 1qpb).

## 2.3. Homo- versus Heterodimerisation

The molecular interaction of subunits in oligomers can be regarded as specific as the formation of the hydrophobic core of a protein. The structure of a protein or subunit determines the kinetic and thermodynamic characteristics of its interaction with other molecules. Therefore, a competition between homo- and heterodimerzation is only expected if the different subunits of the oligomer are highly homologous.

The most prominent example of such a competition is the reconstitution of heterodimeric bacterial luciferase from Vibrio harveyi. The two subunits consisting of a $(\beta\alpha)_8$ barrel are structurally homologous and probably result from gene duplication (35). The association is mediated by two helices from each subunit forming a four-helix bundle at the contact site (36,37). The reconstitution of luciferase from the urea-denatured state is fully reversible. It comprises as rate limiting steps the folding of the subunits on the monomeric level, the association of the subunits to an inactive heterodimer (at low protein concentrations), and a conformational rearrangement on the dimeric level leading to the native state (38-41). As shown by far UV circular dichroism, the inactive dimer seems to be incompletely folded.

Whereas the reconstitution of luciferase is quantitative if the subunits are refolded together, refolding of the subunits separately with a subsequent assembly does not lead to active heterodimer (38,42). In a detailed study it could be shown that active heterodimeric luciferase can be assembled upon folding of the β-subunit in the presence of the pre-folded α-subunit, but not vice versa (43). In the absence of the α-subunit, the β-subunit forms a homodimer which is almost as stable thermodynamically as the heterodimer. Furthermore, the $\beta_2$-homodimer is kinetically very stable, and under native conditions the dissociation does not occur on a biological or experimental timescale (43).

Upon reconstitution, the association of the $\beta_2$-homodimer and the αβ-heterodimer compete with each other. The reason of the almost exclusive formation of the heterodimer is due to the kinetics of association: The rate of heterodimerization is more than tenfold higher than that of the $\beta_2$-homodimerization. The structures of the homo- and heterodimer are almost identical regarding the contacts within the interfaces; most of the inter-subunit hydrogen bonds and water-mediated contacts are conserved between the two dimers (36,37). Thus from the structural point of view there is no obvious explanation for the different association kinetics of homo- and heterodimeric luciferase.

As another example for the competition of homo- and hetero-oligomerization may serve mammalian lactate dehydrogenase (LDH). LDH is a tetrameric protein and exists in two homomeric variants in skeletal muscle (LDH-$M_4$) and in the heart (LDH-$H_4$). If these two isoforms are reconstituted together from the denatured monomers, the product is not only the two homomeric tetramers $M_4$ and $H_4$, but all types of hybrids can be observed. Even more surprising the distribution of all the variants fits to a statistical process, e.g. the ratio of $M_4$, $M_3H_1$, $M_2H_2$, $M_1H_3$, and $H_4$ is 1:4:6:4:1 (38,9). The beauty of this system is that it can be used to analyse the two different association steps upon reconstitution; at different time points of the reconstitution of LDH-$M_4$ an excess of LDH-H subunits can be added to the reaction. From the ratios of the hybrids reconstituted after addition of LDH-H subunits to the refolding LDH-M, the amount of monomeric, dimeric, and tetrameric M-subunits can be calculated (38).

## 2.4. Stability of Oligomeric Proteins

The laws of thermodynamics require that association equilibria are shifted towards the dissociated state at low protein concentrations. And indeed, for several oligomers it has been shown that the dissociation constants of dimerization are in the nanomolar to micromolar range, and thus the dissociation can be achieved directly by dilution experiments. Examples are phosphofructokinase (45), dimeric Arc repressor (46), Hsp33 (47), and yeast hexokinase (48,49). In some cases this association equilibrium is functionally and physiologically relevant. Dimeric Hsp33 is active as a molecular chaperone, whereas the monomeric form is less or non-active (47,50). The $K_D$ describing the monomer dimer equilibrium is in the range of the physiological concentration of Hsp33 in *E.coli* (47). In the case of yeast hexokinase, the monomeric state of the protein is less active than the dimeric form. Interestingly, the dimerization is favoured in the presence of the substrate glucose (48).

Most oligomeric proteins, however, dissociate only at concentrations too low to observe dissociation by mere dilution (9). This finding suggests that subunit interactions account for a large fraction of the conformational stability of oligomeric proteins; sub-nanomolar dissociation equilibrium constants for dimerization correspond to standard free energy changes of > 50 kJ/mol. In such cases, dissociation is only achievable by denaturation. In the most simple case of a reversibly denaturing dimeric protein this transition can be ascribed to the reaction $N_2$ <-> 2U (51). Quantitative analysis of these data yields the free energy of stabilisation of the native dimer which comprises the free energy of the folded monomer and the free energy of the subunit association. Oligomeric states are often not only thermodynamically stable, but they are also kinetically stabilized, that is the activation energy of dissociation under native conditions is very high. Dissociation rates in denaturant-free buffer have been estimated by extrapolation to be as low as $10^{-14}$ $s^{-1}$ (52).

## 2.5. Kinetics of Folding and Association

### General Considerations

The reconstitution of oligomeric proteins from the denatured state or during biosynthesis generally constitutes a series of uni- and bimolecular steps. For a dimeric protein reconstitution involves the following steps:

$$2U \rightarrow 2M \rightarrow M_2 \rightarrow N_2$$

where U is the denatured state. M represents a folded monomer, $M_2$ a non-native dimer, and $N_2$ is the native dimer. The folding reactions preceding subunit assembly are identical to those of single-domain or multi-domain proteins. They consist of a hydrophobic collapse, formation of secondary structure, and the merging of domains, leading to association-competent monomers. As in the case of monomeric proteins these reactions may occur as sequential or parallel reactions on multiple pathways (53,54). All these reactions follow first-order kinetics and do not depend on protein concentration. In contrast, the association of subunits (2M -> $M_2$) is a bimolecular and therefore probably a second order reaction. Subsequently, the dimer may need to undergo further unimolecular rearrangements to reach the final native state ($M_2$ -> $N_2$). Because the half-time of association is concentration dependent, the nature of the rate-limiting step of reconstitution of oligomeric proteins may change with protein concentration (9). At high protein concentration the formation of association-competent subunits is rate limiting, at low protein concentration dimerisation is rate-limiting, and the overall reaction follows second order kinetics. Under these conditions often highly structured, association-competent subunits are populated kinetically during refolding/assembly. In general, such intermediates possess a native-like secondary structural content that does not change significantly upon subsequent association. Thus upon folding/association the change of secondary structure as analysed e.g. by CD spectroscopy follows a comparably fast first order kinetics reaching the signal of the native protein as subunit folding is completed. In contrast, if reconstitution is analysed by reactivation, that is regain of enzyme activity, no reactivation is observed during subunit folding provided that the non-associated subunits do not possess enzyme activity by themselves. The folding of the subunits results in a lag-phase in regain in activity; reactivation occurs upon association. Thus, the overall reactivation is described by a uni-bi-molecular process which follows a sigmoidal time course, with the lag-phase determined by first-order subunit folding (Fig. 3) (55). In some cases the folded but non-associated subunit already show ligand binding or partial enzyme activity. Examples are mammalian aldolases (56), lactate dehydrogenase (57), the P22 tailspike protein (58), and yeast hexokinases (48,49).

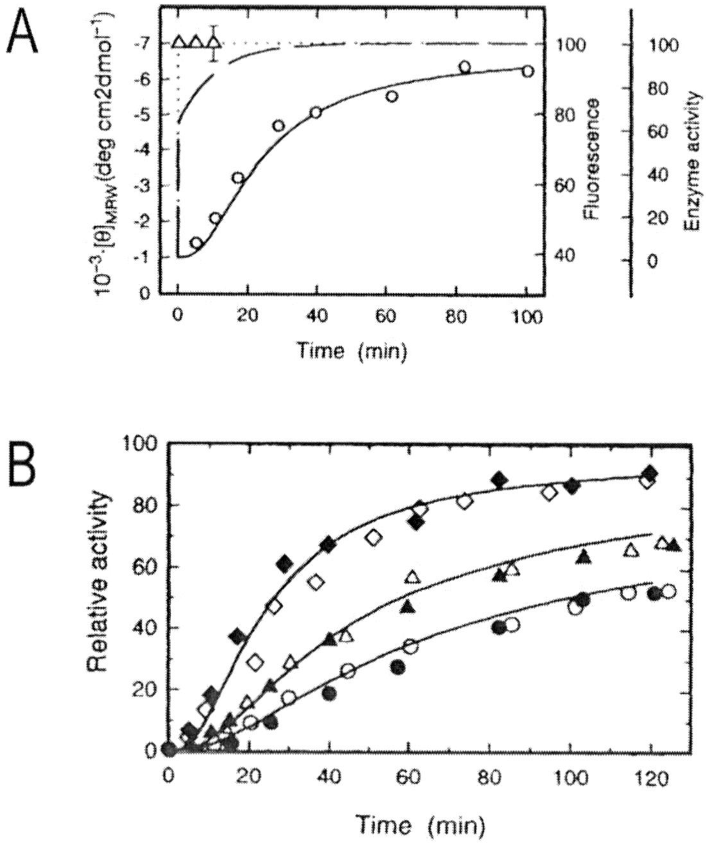


Figure 3. Folding and association pathway of yeast invertase. (A) Changes in circular dichroism (open triangles), fluorescence emission (broken line), and enzymatic activity (open circles) occur with very different kinetics during the reconstitution of homodimeric yeast invertase. (B) The first-order folding rate obtained from the fluorescence change can be used to fit a sequential uni-bimolecular mode to the reactivation data obtained at different protein concentrations (full lines). Subunit concentrations were 8.5 nM (circles), 17 nM (triangles) and 68 nM (diamonds) (55).

The assembly of larger oligomers can be described accordingly as a series of multiple uni- and bimolecular steps. The assembly of a homotrimer is expected to proceed via a dimeric intermediate

$$3M <-> M_2 + M -> M_3$$

In this reaction scheme, the first association resulting in a dimer, in which only one of two interfaces is buried, is considered as a fast equilibrium. This first association step is unfavourable and the dimeric intermediate is not populated in equilibrium. The second association step will pull this equilibrium towards the native trimeric state. In this concept, the overall process follows apparent third-order kinetics (59).

A tetramer may be envisioned as a dimer of dimers. Thus, the scheme for this reaction is just an extension of that shown for a dimeric protein (9, 60).

$$4U \rightarrow 4M \rightarrow 2M'_2 \rightarrow 2M_2 \rightarrow M_4 \rightarrow N_4$$

In the case of multimers such as fibrillary proteins or virus capsids the most simple model comprises a stepwise addition of monomers to the growing chain.

$$2A \leftrightarrow A_2,\ A_2 + A \leftrightarrow A_3,\ A_3 + A \leftrightarrow A_4,\ \ldots$$

In this scheme the equilibrium constants of association may be all identical. However, often the first steps of multimerisation are thermodynamically disfavoured, that is the equilibrium is on the side of the non-associated species. Only after reaching a certain size of the oligomer structure, further addition of monomers to this nucleus becomes thermodynamically favourable. This would lead to an overall assembly process characterized by a nucleation/polymerisation reaction.

**Table 1. Rates of association of some oligomeric proteins**

| Protein | Subunit mass (kDa) | Reaction | k ($M^{-1}s^{-1}$) | ref. |
|---|---|---|---|---|
| Triosephosphate isomerase | 27 | 2M -> D | $3 \times 10^5$ | 9 |
| Malate dehydrogenase (mitochondrial) | 33 | 2M -> D | $3 \times 10^4$ | 9 |
| Invertase | 76 | 2M -> D | $1 \times 10^4$ | 55 |
| β-Galactosidase | 116 | 2M -> D | $4 \times 10^3$ | 97 |
| Alcohol dehydrogenase (liver) | 40 | 2M -> D | $2 \times 10^3$ | 9 |
| Phosphoglycerate mutase | 27 | 2M -> D | $6 \times 10^3$ | 98 |
| | | 2D -> T | $3 \times 10^4$ | 98 |
| lactate dehydrogenase | 36 | 2D -> T | $2 \times 10^4$ | 9 |
| Arc repressor | 62 | 2M -> D | $1 \times 10^7$ | 64 |
| Trp aporepressor | 12 | 2M -> D | $3 \times 10^8$ | 63 |

*Rates of association*

The maximum rate of association of two molecules is determined by their random diffusional collision. For a common protein domain this limit is about $10^9\ M^{-1}s^{-1}$ (61). Indeed, this limit is reached in the binding of chaperones to their substrate proteins. Chaperones recognize a wide variety of substrate proteins mainly by hydrophobic interactions. Since these interactions do not need a pre-orientation of chaperone and substrate, most of the diffusional collisions result in complex formation. However, this is quite different for the specific association of folded subunits. Here, the pre-formed interfaces require an exact rotational alignment for association, thus reducing the diffusion-controlled rate by several orders of magnitude. Brownian dynamics simulations estimate the rate of association of model proteins to be $10^6$ $M^{-1}s^{-1}$ (62). Most second order rate constants determined experimentally are even smaller (Table 1). There are, however, two examples published, the core domain of trp repressor and the arc repressor, that show faster dimerization than expected (Table 1, (63,64)). In both cases folding is tightly coupled to association; the association-competent monomers are only partially structured.

# III. Special Motifs Facilitating Oligomer Formation

Most proteins exert their biological functions in homo- or hetero-oligomeric complexes. In some cases, especially in highly regulated signal transduction processes, protein-protein association is mediated by specialised protein domains. Such domains are for example 14-3-3 and WW domains, domains of the type $S_H2$ or $S_H3$, or PDZ and PH domains (65-68). In most cases, these domains mediate only transient, low-affinity interactions, ideally suited for functional regulation e.g. by phosphorylation/dephosphorylation processes. In the following we rather focus on some specific structural aspects conferring a stable association of subunits.

## 3.1. Domain Swapping

In principal subunit interfaces of oligomeric proteins are not different from domain-domain interfaces of multi-domain proteins. Therefore, from a structural point of view it should be possible that a two-domain protein can switch between intramolecular domain-domain interactions in its monomeric state and intermolecular domain-domain interactions in a dimeric or even higher oligomeric state. In an evolutionary context point mutations in the respective protein might then favor either the monomeric or the oligomeric state.

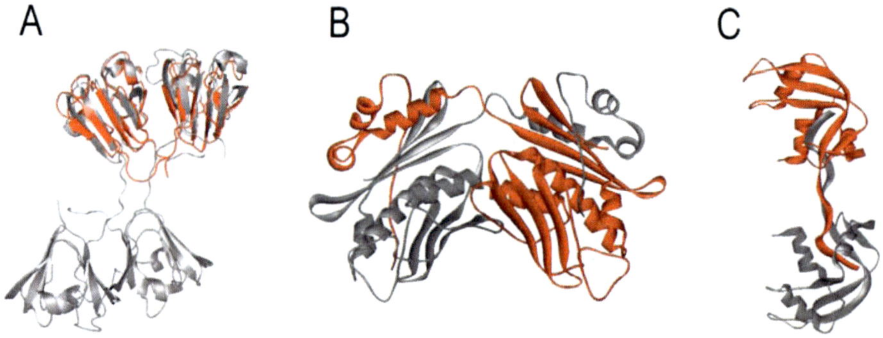


Domain swapping can occur on different levels of the structural hierarchy. (A) β-crystallin dimer (gray) as potential product of swapped domains of γ-crystallin (red). (B) In Hsp33 (pdb: 1hw7) the two monomers swap a supersecondary structure consisting of three helices and two β-strands. (C) Domain swapping in dimeric RNase A (pdb: 1f0v) occurs just on the level of a single β-strand.

Figure 4. Domain swapping.

This concept of domain swapping is illustrated by βγ-crystallins. βγ-crystallins are members of the superfamily of β-sheeted proteins containing the greek key motif. γ-crystallin is monomeric even at extremely high protein concentrations, whereas β-crystallin forms dimers or higher oligomers. Both show highly homologous two-domain structures. Comparison of the structures of γ- and β-crystallins suggests that the β dimers are the product of a domain swapping and association of a γ ancestor (Fig. 4A). This hypothesis is supported by the fact that 60 % of the interface-residues of the N- and C-terminal domains of γ-crystallin are identical or highly conserved compared to those of the N- and C-terminal domains of the two subunits of β-crystallin (69).

Domain swapping does not correlate necessarily with structural domains. Instead, the swapped portion can vary from a large globular domain, as described for βγ-crystallins, to supersecondary structural elements or even a single helix or β-strand. Of the ca. 40 structurally characterized cases of domain-swapped proteins most swapped elements are either at the N- or C-terminus and are diverse in their primary and secondary structure (70). Hsp33, for example, forms a dimer, in which a C-terminal arm of one monomer consisting of two α-helices wraps around a second monomer (Fig. 4B) (71, 72). This C-terminal arm is connected to the remaining part of the protein by a putatively flexible hinge region. The domain swapping of Hsp33 is connected to its function as chaperone. The chaperone-active dimer is in equilibrium with a less active or even inactive monomer ($K_D$ = 0.6 μM, (47)). This equilibrium is controlled by the oxidation state of the protein since Hsp33 is a redox-regulated chaperone (47,73). In another example, the transcription factor NFκBp50 has been shown to be functional only in a dimeric state. Chirgadze et al. analysed a variant of NFκBp50 with a mutation placed in the dimer interface. This protein is still a highly stable dimeric and functional protein; the mode of dimerisation, however, is quite different from that of the wildtype protein. Whereas the wildtype is a "side-by-side" dimer the mutant shows a highly intertwined domain swapping, exchanging three strands of an immunoglobulin-like fold (74).

Whereas in NFκBp50 half of a β-sheet domain is swapped, in case of RNase A the swapped portion in the domain swapped dimer consists only of a single β-strand of 14 residues (Fig. 4C) (75). Even though the interaction site of this dimer is comparably small, the dimer exerts considerable stability. Domain swapping involves passing through an open monomer or partially unfolded structure. Because of this, domain swapped proteins typically do not equilibrate with their closed monomer counterparts at an appreciable rate, unless this equilibrium is triggered by external factors.

Higher association states of domain-swapped proteins are found in several virus capsids, in which capsomers swap β-strands, or pathologic aggregates in the case of α1-antitrypsin (76). Recently, a runaway domain swapping has been shown to facilitate amyloid formation in some cases. T7 endonuclease 1 is structurally a domain swapped dimer. Gou and Eisenberg (77) could demonstrate that partial opening of this dimeric structure can lead to a runaway domain swapping. In this reaction two monomers do not simply exchange the swapped domain but instead each protein molecule swaps the domain into the complementary domain of the adjacent molecule along the amyloidogenic fibril.

## 3.2. Coiled Coils

The coiled coil is a common structural motif, involving 3-5 % of all amino acids in proteins (78,79). It consists of 2 to 5 -helices wrapped around each other into a left-handed supercoil. In this supercoil the helices are slightly distorted that lowers the periodicity of 3.6

residues per complete turn in a usual α-helix to 3.5 residues per turn in the supercoiled helix. Thus, a heptad repeat (denoted a-b-c-d-e-f-g) occurs every two turns in the helix (Fig. 5) (80,81). The PV (peptide Velcro) hypothesis states three structural elements vital to the formation of a specific coiled coil (82). Firstly, the amino acids at positions a and d must be hydrophobic. Secondly, residues e and g must be charged, allowing interhelical electrostatic interactions to be formed. Thirdly, residues at the positions b, c, and f should be hydrophilic as these will form the solvent exposed helical surface. The PV hypothesis is only a general description of coiled coils, subtle variations of the amino acid content at each position determine orientation, specificity, and oligomerisation state of a specific coiled coil.

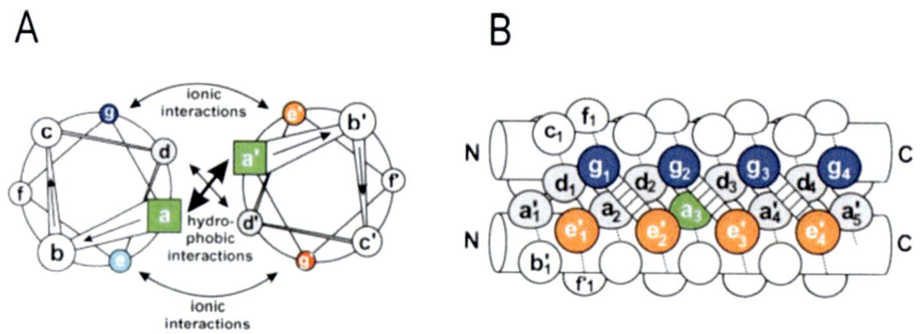


The coiled coil is shown in a top view (A) and a side view (B). The residues of the heptad repeat are labeled a-g and a'-g', respectively. The hydrophobic core is formed by the residues a, a', d, d'. Ionic interactions occur between the pairs e/g' and e'/g (adapted from (78).

Figure 5. Schematic drawing of a parallel dimeric coiled coil.

The hydrophobic a and d repeats facilitate dimerisation along one face of each helix, thus representing the dominating contribution to the overall stability of the coiled coil. Variation in packing environments at these two sites results in different preferences for hydrophobic residues. For example, the yeast transcription factor GCN4 is a parallel homodimeric coiled coil with b-branched side chains in the a position while position d favors a g-branched leucine (83). In the GCN4 mutant p-LI, where leucine is located in position a whereas position d is occupied by isoleucine, a tetramer is formed (84). Despite the hydrophobic nature of the a/d positions, often a few polar residues are introduced at these sites conferring specificity of the coiled coil formation. In GCN4 an asparagine is located at one a position mediating dimmer formation through inter-helical hydrogen bonding as observed by NMR and x-ray studies (85).

Pairing specificity of the helices within the coiled coil is greatly influenced by electrostatic interactions of residues in position e and g in the heptad repeat. Charge pattern on the surface of the coiled coil control the preference for homo- or heterotypic pairing in an parallel or antiparallel orientation (82,86,87). In a three-helix coiled coil the e an g positions also contain apolar residues because the solvent excluded surface in a three-helix coil is smaller than in a two-helix coil.

In several coiled coil proteins a thirteen-residue sequence has been found to be required for initiation of the coiled coil assembly, suggesting the existence of an autonomous helical folding unit acting as a scaffold around which the remainder of the coiled coil can assemble

(88). However, many designed coiled coils do not possess such an initiation sequence indicating that such a helical nucleus is not essential.

Besides the left-handed coiled coil described before, very few examples are known to possess a right-handed coiled coil. In these coiled coils the number of residues per turn is 3.67 (eleven residues every three turns of a helix). The hydrophobic packing in these right-handed coiled coils is quite different from that of the felt-handed counterparts and the core contains water molecules (89).

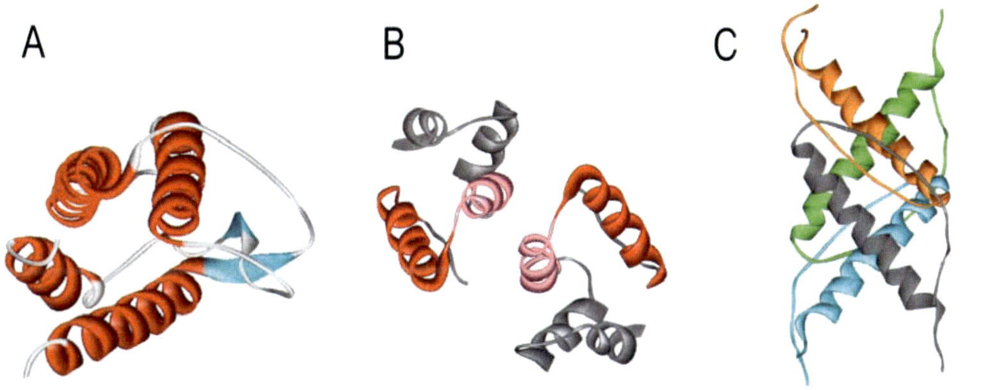


(A) Monomeric interleukin 4 (pdb: 1itl) is a single domain protein consisting just of a four-helix bundle. (B) The C-terminal domain of HIV-1 capsid protein (CA146-231, pdb: 1aum) forms a four-helix bundle where each subunit contributes a helix-loop-helix motif (marked in red). (C) In the tetramerisation domain of p53 (pdb: 1AIE) the four helices of the bundle are provided by four different subunits. The arrangement of the four helices is depicted.

Figure 6. Four-helix bundles in context of different oilgomeric states.

## 3.3. Four-Helix-Bundles

The four-helix-bundle is an often found structural motif in proteins. It is based on amphiphilic helices that contain a pattern of polar and nonpolar residues reflecting the periodicity of α-helices of 3.6 residues per complete turn (e.g. polar-polar-nonpolar-polar-polar-nonpolar-nonpolar). The motif can be classified into different folds regarding topological and geometrical properties (90). Topological variables are the orientation of the helices in an parallel or antiparallel fashion can the connectivity of the helices. Chothia and coworkers described the interaction of the helices by a "ridges-into-grooves" model (91,92). According to this model the side chains projecting out of the helicesgenerate ridges on their surfaces. Two sets of such ridges are formed, one formed by every three residues (i-3, i, i+3,...) and the other by every four residues. If the i+4 ridges of one helix intercalate into the i+4 grooves of another helix, the respective helices will cross each other at an angle of –50°. On the other hand, if i+4 ridges of one helix intercalate into the i+3 grooves of the other helix the crossing will be 20°. Due to these crossing angles the helices in a four helix bundle will splay apart from each other unless this structural consequence is prohibited by a helical kink or a supercoiling of the helices (93). In the latter case the structure resembles a coiled coil discussed before.

Four helix bundles can be found as isolated fold in monomeric proteins such as cytokines (Fig. 6A) (94), as domains in larger monomeric proteins (e.g. T4 lysozyme), and as assembly domain in protein oligomers. Such an oligomerisation doamin will lead to dimer formation if each monomer contributes two helices of the four helix bundle as has been shown for the C-terminal dimerisation domain of the HIV-1 capsid protein (Fig. 6B) (95). On the other hand, tetramerisation will be induced if all four helices belong to different subunits. Such a structure is observed in the tetramerisation domain of p53. In this domain a monomer consisiting of a β-strand and an α-helix associates with a second monomer across an antiparallel β-sheet and an antiparallel helix-helix interface to form a dimer. Two of these dimers associate via parallel helix-helix interfaces to the tetramer (Fig. 6C) (96).

## Conclusions

Most proteins exist as homo- or heterooligomeric complexes in their functional and biological context. The molecular forces controlling these association reactions are the same as those determining the structure of monomeric proteins and intramolecular domain-domain interactions. Nevertheless, a remarkable diversity of subunit interfaces can be found, especially regarding the spatial arrangement of hydrophobic and polar residues as well as the position of water molecules. An mechanism of dimerisation and oligomerisation of increasingly importance is the domain swapping. Here single secondary structural elements up to complete domains can be swapped between different protein molecules, thus leading to dimers, oligomers, and even amyloid fibrils. Furthermore, protein motifs virtually designed for the purpose of complex formation are found in helical structures such as coiled coils and four-helix bundles.

## References

[1] Lander, E.S., Linton, L.M., Birren, B., Nusbaum, C., Zody, M.C., Baldwin, J., Devon, K., Dewar, K., Doyle, M., FitzHugh, W. et al. (2001). Initial sequencing and analysis of the human genome. *Nature,* **409**, 860-921.

[2] Venter, J.C., Adams, M.D., Myers, E.W., Li, P.W., Mural, R.J., Sutton, G.G., Smith, H.O, Yandell, M., Evans, C.A., Holt, R.A. et al. (2001). The sequence of the human genome. *Science,* **291**, 1304-1351.

[3] Alberts, B. (1998). The cell as a collection of protein machines: preparing the next generation of molecular biologists. *Cell,* **92**, 291-294.

[4] Kumar, A. & Snyder, M. (2002). Protein complexes take the bait. *Nature,* **415**, 123-124.

[5] von Mering, C., Krause, R., Snel, B., Cornell, M., Oliver, S.G., Fields, S. & Bork, P. (2002) Comparative assessment of large-scale data sets of protein-protein interactions. *Nature,* **417**, 399-403.

[6] Lawson, D.M., Artymiuk, P.J., Yewdall, S.J., Smith, I.M., Livingstone, I.C., Treffry, A., Luzzago, A., Levi, S., Arosio, P., Cesareni, et al. (1991). Solving the structure of human H ferritin by genetically engineering intermolecular crystal contacts. *Nature,* **349**, 541-544.

[7] Xu, Z., Horwich, A.L. & Sigler, P.B. (1997). The crystal structure of the asymmetric GroEL-GroES-( ADP)7 chaperonin complex. *Nature*, **388**, 741-750.
[8] Izard, T., Aevarsson A., Allen, M.D., Westphal, A.H., Perham, R.N., de Kok, A., & Hol, W. G. (1999). Principles of quasi-equivalence and euclidean geometry govern the assembly of cubic and dodecahedral cores of pyruvate dehydrogenase complexes. *Proc. Natl Acad. Sci. USA*, **96**, 1240-1245.
[9] Jaenicke, R. (1987). Folding and association of proteins. *Frog. Biophys. Mol. Biol.*, **49**, 117-237.
[10] Gerstein, M. & Krebs, W. (1998). A database of macromolecular motions. *Nucleic Acids Res*, **26**, 4280-4290.
[11] Echols, N., Milburn, D. & Gerstein, M. (2003). MolMovDB: analysis and visualization of conformational change and structural flexibility. *Nucleic Acids Res*, **31**, 478-482.
[12] Freire, E. (1999). The propagation of binding interactions to remote sites in proteins: analysis of the binding of the monoclonal antibody D1.3 to lysozyme. *Proc Natl Acad Sci USA*, **96**, 10118-10122.
[13] Seckler, R. (2000). Assembly of multi-subunit strucutres. In: Mechanisms of Protein Folding, Ed.: R.H. Pain, 2nd ed., Oxford Press, pp. 279-308.
[14] Weber, G., Schorgendorfer, K., Schneider-Scherzer, E., & Leitner, E. (1994). The peptide synthetase catalyzing cyclosporin production in Tolypocladium niveum is encoded by a giant 45.8-kilobase open reading frame. *Curr. Genet.*, **26**, 120-125.
[15] Dautry-Varsat, A. & Garel, J.R. (1978). Refolding of a bifunctional enzyme and its monofunctional fragment. *Proc. Natl Acad. Sci. USA*, **75**, 5979-5982.
[16] Creighton, T.E. (1992). Proteins: structures and molecular properties, (2nd edn). W.H. Freeman, New York, p. 309-325.
[17] Miller, S., Schuler, B., & Seckler, R. (1998). Phage P22 tailspike protein: removal of head binding domain unmasks effects of folding mutations on native-state thermal stability. *Prot. Sci.*, **7**, 2223-2232.
[18] Danner, M., Fuchs, A., Miller, S., and Seckler, R. (1993). Folding and assembly of phage P22 tailspike endorhamnosidase lacking the N-terminal, head-binding domain. *Eur. J. Biochem.*, **215**, 653-661.
[19] Jones, S. & Thornton, J.M. (1997). Analysis of protein protein interaction sites using surface patches. *J. Mol. Biol.*, **272**, 121-132.
[20] Tsai, C.J., Lin, S.L., Wolfson, H.J. & Nussinov, R. (1997). Studies of protein-protein interfaces: a statistical analysis of the hydrophobic effect. *Protein Sci.*, **6**, 53-64.
[21] Tsai, C.J., Xu, D. & Nussinov, R. (1997). Structural motifs at protein-protein interfaces: protein cores versus two-state and three-state model complexes. *Prot. Sci.*, **6**, 1793-1805.
[22] Le Conte, L., Chothia, C. & Ianin, J. (1999). The atomic structure of protein-protein recognition sites. *J. Mol. Biol.*, **285**, 2177-2198.
[23] Chothia, C. & Janin, J. (1975). Principles of protein-protein recognition. *Nature*, **256**, 705-708.
[24] Dill, K.A. (1990). Dominant forces in protein folding. *Biochemistry*, **29**, 7133-7155.
[25] DeLano, W.L. (2002). Unravelling hot spots in the binding interface: progress and challenges. *Curr. Opin. Struct. Biol.*, **12**, 14-20.
[26] Larsen, T.A., Olson, A.J. & Goodsell, D.S. (1998). Morphology of protein-protein interfaces. *Structure*, **6**, 421-427.

[27] Hubbard, S.J. & Argos, P. (1994). Cavities and packing at protein interfaces. *Protein Sci.* **3**, 2194–2206.

[28] Uversky, V.N. (2002). Natively unfolded proteins: A point where biology waits for physics. *Prot. Science*, **11**, 739-756.

[29] Venyaminov, S. Yu., Gudkov, A.T., Gogia, Z.V. & Tumanova, L.G. (1981). Absorption and cicular dichroism spectra of individual proteins from Escherichia coli ribosomes. Pushkino, Russia.

[30] Yusupov, M.M., Yusupova, G.Z., Baucom. A., Lieberman, K., Earnest, T.N., Cate, J.H. & Noller, H.F. (2001). *Science,* **292**, 883-896.

[31] Kriwacki, R.W., Hengst, L., Tennant, L., Reed, S.I. & Wright P.E. (1996). Structural studies of p21 in the free and Cdk2-bound state: Conformational disorder mediates binding diversity. *Prog. Natl Acad. Sci. USA*, **93**, 11504-11509.

[32] Harris, B.Z., Venkatasubrahmanyam, S. & Lim, W.A. (2002) Coordinated folding and association of the LIN-2, -7 (L27) domain. *J. Biol. Chem.*, **277**, 34902-34906.

[33] Powers, E.T. & Powers, D.L. (2004). A perspective of mechanisms of protein tetramer formation. *Biophys. J.,* **85,** 3587-3599.

[34] Killenberg-Jabs, M., Jabs, A., Lilie, H., Golbik, R. & Hubner, G. (2001). Active oligomeric states of pyruvate decarboxylase and their functional characterization. *Eur. J. Biochem.,* **268,** 1698-1704.

[35] Fisher, A.J., Raushel, F.M., Baldwin, T.O. & Rayment, I. (1995). Three-dimensional structure of bacterial luciferase from Vibrio harveyi at 2.4 A resolution. *Biochemistry*, **34**, 6581-6586.

[36] Thoden, J.B., Holden, H.M, Fisher, A.J., Sinclair, J.F., Wesenberg, G., Baldwin, T.O. & Rayment, I. (1997). Structure of the beta 2 homodimer of bacterial luciferase from Vibrio harveyi: X-ray analysis of a kinetic protein folding trap. *Protein Sci.*, **6**, 13-23.

[37] Tanner, J.J., Miller, M.D., Wilson, K.S., Tu, S.C. & Krause, K.L. (1997). Structure of bacterial luciferase beta 2 homodimer: implications for flavin binding. *Biochemistry*, **36**, 665-672.

[38] Waddle, J.J., Johnston, T.C. & Baldwin, T.O. (1987). Polypeptide folding and dimerization in bacterial luciferase occur by a concerted mechanism in vivo. *Biochemistry*, **26**, 4917-4921.

[39] Ziegler, M.M., Goldberg, M.E., Chaffotte, A.F. & Baldwin, T.O. (1993). Refolding of luciferase subunits from urea and assembly of the active heterodimer. Evidence for folding intermediates that precede and follow the dimerization step on the pathway to the active form of the enzyme. *J. Biol. Chem.*, **268**, 10760-10765.

[40] Baldwin, T.O., Ziegler, M.M., Chaffotte, A.F. & Goldberg, M. E. (1993). Contribution of folding steps involving the individual subunits of bacterial luciferase to the assembly of the active heterodimeric enzyme. *J. Biol. Chem.*, **268**, 10766-10772.

[41] Clark, A.C., Sinclair, J.F. & Baldwin, T.O. (1993). Folding of bacterial luciferase involves a non-native heterodimeric intermediate in equilibrium with the native enzyme and the unfolded subunits. *J. Biol. Chem.*, **268**, 10773-10779.

[42] Sinclair, J.F., Waddle, J.J., Waddill, E.F. & Baldwin, T.O. (1993). Purified native subunits of bacterial luciferase are active in the bioluminescence reaction but fail to assemble into the alpha beta structure. *Biochemistry*, **32**, 5036-5044.

[43] Sinclair, J.F., Ziegler, M.M. & Baldwin, T.O. (1994). Kinetic partitioning during protein folding yields multiple native states. *Nature Struct. Biol.,* **1**, 320-326.

[44] Hermann, R., Rudolph, R. & Jaenicke, R. (1982). The use of subunit hybridization to monitor the reassociation of porcine lactate dehydrogenase after acid dissociation. *Hoppe Seylers Z. Physiol. Chem.*, **363**, 1259-1265.

[45] Kono, N., Uyeda, K. & Oliver, R.M. (1973). Chicken liver phosphofructokinase. I. Crystallization and physicochemical properties. *J. Biol. Chem.*, **248**, 8592-8602.

[46] Bowie, J.U. & Sauer, R.T. (1989). Equilibrium dissociation and unfolding of the Arc repressor dimer. *Biochemistry*, **28**, 7139-7143.

[47] Graumann, J., Lilie, H., Tang, X., Tucker, K.A., Hoffmann, J., Janakiraman, V., Saper, M., Bardwell, J.C.A. & Jakob, U. (2001). Activation of the redox regulated chaperone Hsp33 - a two step mechanism. *Structure,* **9**, 377-387.

[48] Golbik, R., Naumann, M., Otto, A., Müller, E.-C., Behlke, J., Reuter, R., Hübner, G. & Kriegel, T.M. (2001). Regulation of phosphotransferase activity of hexokinase 2 from Saccharomyces cerevisiae by modification at serine-14. *Biochemistry,* **40,** 1083–1090.

[49] Baer, D., Golbik, R., Huebner, G., Lilie, H., Mueller, E.C., Naumann, M., Otto, A., Reuter, R., Breunig, K.D. & Kriegel, T.M. (2003). The unique hexokinase of Kluyveromyces lactis: Isolation, molecular characterization and evaluation of a role in glucose signaling. *J. Biol. Chem.,* **278**, 39280-39286.

[50] Hoffmann, J.H., Linke, K., Graf, P.C., Lilie, H. & Jakob, U. (2004). Identification of a redox regulated chaperone network. *EMBO J.*, **23,** 160-168.

[51] Neet, K.E. & Timm, D.E. (1994). Conformational stability of dimeric proteins: quantitative studies by equilibrium denaturation. *Protein Sci.*, **3**, 2167-2174.

[52] Sinclair, J.F., Ziegler, M.M. & Baldwin, T.O. (1994). Kinetic partitioning during protein folding yields multiple native states. *Nature Struct. Biol.,* **1**, 320-326.

[53] Weissman, J.S. (1995). All roads lead to Rome? The multiple pathways of protein folding. *Chem Biol.,* **2**, 255-260.

[54] Dill, K.A. & Chan, H.S. (1997). From Levinthal to pathways to funnels. *Nat. Struct. Biol.,* **4,** 10-19.

[55] Kern, G., Kern, D., Jaenicke, R. & Seckler, R. (1993). Kinetics of folding and association of differently glycosylated variants of invertase from Saccharomyces cerevisiae. *Prot. Science*, **2**, 1862-1867.

[56] Rudolph, R., Westhof, E. & Jaenicke, R. (1977). Kinetic analysis of the reactivation of rabbit muscle aldolase after denaturation with guanidine-HCl. *FEBS Lett.*, **73**, 204-206.

[57] Opitz, U., Rudolph, R., Jaenicke, R., Ericsson, L. & Neurath, H. (1987). Proteolytic dimers of porcine muscle lactate dehydrogenase: characterization, folding, and reconstitution of the truncated and nicked polypeptide chain. *Biochemistry*, **26,** 1399-13406.

[58] Miller, S., Schuler, B. & Seckler, R. (1998). A reversibly unfolding fragment of P22 tailspike protein with native structure: the isolated beta-helix domain. *Biochemistry*, **37,** 9160-9168.

[59] Burns, D.L. & Schachman, H.K. (1982). Assembly of the catalytic trimers of aspartate transcarbamoylase from folded monomers. *J. Biol. Chem.*, **257**, 8638-8647.

[60] Jaenicke, R. & Lilie, H. (2000). Folding and association of oligomeric and multimeric proteins. *Adv. Prot. Chem.,* **53**, 329-401.

[61] Koren, R. & Hammes, G.G. (1976). A kinetic study of protein-protein interactions. *Biochemistry*, **15**, 1165-1171.

[62] Northrup, S.H. & Erickson, H.P. (1992). Kinetics of protein-protein association explained by Brownian dynamics computer simulation. *Proc. Natl Acad. Sci., USA*, **89**, 3338-3342.

[63] Gloss, L.M. & Matthews, C.R. (1998). Mechanism of folding of the dimeric core domain of Escherichia coli trp repressor: a nearly diffusion-limited reaction leads to the formation of an on-pathway dimeric intermediate. *Biochemistry*, **37**,15990-15998.

[64] Milla, M.E. & Sauer, R.T. (1994). P22 Arc repressor: folding kinetics of a single-domain, dimeric protein. *Biochemistry*, **33**, 1125-1131.

[65] Espejo, A., Cote, J., Bednarek, A., Richard, S. & Redford, M.T. (2002). A protein-domain microarray identifies novel protein-protein interactions. *Biochem J.* **367**, 697-702.

[66] Boggon, T.J. & Eck, M.J. (2004). Structure and regulation of Src family kinases. *Oncogene,* **23**, 7918-27.

[67] Kay, B.K., Williamson, M.P. & Sudol, M. (2000). The importance of being proline: the interaction of proline-rich motifs in signaling proteins with their cognate domains. *FASEB J.* **14**, 231-41.

[68] Birge, R.B., Knudsen, B.S., Besser, D. & Hanafusa, H. (1996). SH2 and SH3-containing adaptor proteins: redundant or independent mediators of intracellular signal transduction. *Genes Cells*, **1**, 595-613.

[69] Bax, B., Lapatto, R., Nalini, V., Driessen, H., Lindley, P.V., Mahadevan, D., Blundell, T.L. & Slingsby, C. (1990). X-ray analysis of beta B2-crystallin and evolution of oligomeric lens proteins. *Nature*, **347**, 776-779.

[70] Liu, Y., & Eisenberg, D. (2202). 3D domain swapping: As domains continue to swap. *Prot. Science*, **11**, 1285-1299.

[71] Vijayalakshmi, J., Mukhergee, M.K., Graumann, J., Jakob, U. & Saper, M.A. (2001). The 2.2 A crystal structure of Hsp33: A heat shock protein with a redox-regulated chaperone activity. *Structure,* **9**, 367-375.

[72] Kim, S.J., Jeong, D.G., Chi, S.W., Lee, J.S. & Rya, S.E. (2001). Crystal structure of proteolytic fragments of the redox-sensitive Hsp33 with constitutive chaperone activity. *Nat. Struct. Biol.*, **8,** 459-466.

[73] Graf, P.C., Martinez-Yamout, M., Van Haerents, S., Lilie, H., Dyson, J.H. & Jakob, U. (2004) Activation of the Redox Regulated Chaperone Hsp33 by Domain Unfolding. *J. Biol. Chem.,* **279,** 20529-20538.

[74] Chirgadze, D.Y., Demydschuk, M., Becker, M., Moran, S. & Paoli, M. (2004). Snapshot of protein structure evolution reveals conservation of functional dimerization through intertwined folding. *Structure*, **12**, 1489-1494.

[75] Crestfield, A.M., Stein, W.H. & Moore, S. (1962). On the aggregation of bovine pancreatic ribonuclease. *Arch. Biophys. Biochem. Suppl.* **1**, 217-222.

[76] Lomas, D.A., Evans, D.L., Finch, J.T. & Carrell, R.W. (1992). The mechanism of Z alpha 1-antitrypsin accumulation in the liver. *Nature,* **357,** 541-542.

[77] Guo, Z. & Eisenberg, D. (2006). Runaway domain swapping in amyloid-like fibrils of T7 endonuclease 1. *Proc. Natl. Acad. Sci. USA* **103,** 8042-8047.

[78] Mason, J.M. & Arndt, K.M. (2004). Coiled coil domains: Stability, Specificity, and biological implications. *Chembiochem* **5**, 170-176.

[79] Wolf, E., Kim, P. S. & Berger, B. (1997). MultiCoil: a program for predicting two- and three-stranded coiled coils. *Protein Sci.* **6**, 1179-1189.

[80] Landschulz, W. H., Johnson, P. F. & McKnight, S. L. (1988). The leucine zipper: a hypothetical structure common to a new class of DNA binding proteins. *Science*. **240**, 1759-1764.

[81] Lupas, A. (1996). Coiled coils: new structures and new functions. *Trends Biochem. Sci.* **21**, 375-382.

[82] O'Shea, E. K., Lumb, K. J., Kim, P. S. (1993). Peptide 'Velcro': design of a heterodimeric coiled coil. *Curr. Biol.* **3**, 658-667.

[83] O'Shea, E. K. , Klemm, J. D. , Kim, P. S. & Alber, T. (1991). X-ray structure of the GCN4 leucine zipper, a two-stranded, parallel coiled coil. *Science* **254**, 539-544.

[84] Harbury, P. B. , Zhang, T., Kim, P. S. & Alber, T. (1993). A switch between two-, three- and four-stranded coiled coils in GCN4 leucine zipper mutants. *Science* **262**, 1401-1407.

[85] Junius, F. K. , Mackay, J. P., Bubb, W. A., Jensen, S. A., Weiss, A. S. & King, G. F. (1995). Nuclear magnetic resonance characterization of the Jun leucine zipper domain: unusual properties of coiled-coil interfacial polar residues. *Biochemistry* **34**, 6164-6174.

[86] Graddis, T. J. , Myszka, D. G., & Chaiken, I. M. (1993). Controlled formation of model homo- and heterodimer coiled coil polypeptides. *Biochemistry* **32**, 12664-12671.

[87] Kohn, W. D. , Kay, C. M. & Hodges, R. S. (1995). Protein destabilization by electrostatic repulsions in the two-stranded alpha-helical coiled-coil/leucine zipper. *Protein Sci.* **4**, 237-250.

[88] Kammerer, R. A. , Schulthess, T. , Landwehr, R., Lustig, A., Engel, J. , Aebi, U. & Steinmetz, M. O. (1998). An autonomous folding unit mediates the assembly of two-stranded coiled coils. *Proc. Natl. Acad. Sci. USA* **95**, 13419-13424.

[89] Stetefeld, J., Jenny, M., Schulthess, T., Landwehr, R., Engel, J. & Kammerer, R.A. (2000). Crystal structure of a naturally occurring parallel right-handed coiled coil tetramer. *Nat. Struct. Biol.* **7**, 772-776.

[90] Kamtekar, S. & Hecht, M.H. (1995). The four-helix bundle: what determines a fold? *FASEB J.* **9**, 1013-1022.

[91] Chothia, C., Levitt, M., & Richardson, D. (1981). Helix to helix packing in proteins. *J. Mol. Biol.* **145**, 215-250.

[92] Chothia, C. (1984). Principles that determine the structures of proteins. *Annu. Rev. Biochem.* **53**, 537-572.

[93] Cohen, C. & Parry, D. A. D. (1990). α–Helical coiled coils and bundles: how to design an a-helical protein. *Protein Structure Function Genet.* **7**, 1-15.

[94] Nicola, N.A. & Hilton, D.J. (1998). General classes and functions of four-helix bundle cytokines. *Adv. Prot. Chem.*, **52**, 1-65.

[95] Gamble, T.R., Yoo, S., Vajdos, F.F., von Schwedler, U.K., Worthylake, D.K., Wang, H., McCutcheon, J.P., Sundquist, W.I. & Hill, C.P. (1997). Structure of the carboxyl-terminal dimerization domain of the HIV-1 capsid protein. *Science,* **278**, 849-53.

[96] Jeffrery, P.D., Gorina, S. & Pavletich, N.P. (1995). Crystal structure of the tetramerization domain of the p53 tumor suppressor at 1.7 angstroms. *Science*, **267**, 1498-1502.

[97] Nichtl, A., Buchner, J., Jaenicke, R., Rudolph, R. & Scheibel, T. (1998). Folding and association of beta-galactosidase. *J. Mol. Biol.*, **282**, 1083-1091.

[98] Herrmann, R., Rudolph, R., Jaenicke, R., Price, N.C. & Scobbie, A. (1983). The reconstitution of denatured phosphoglycerate mutase. *J. Biol. Chem.*, **258**, 11014-11019.

In: Protein Folding-Misfolding…
Editors: J.P. Zbilut and T. Scheibel, pp. 55-92

ISBN: 1-60021-417-7


*Chapter 3*

# UNDERSTANDING PROTEIN NON-FOLDING

***Vladimir N. Uversky*[1,2,3,*] *and A. Keith Dunker*[1,2]**

[1]Center for Computational Biology and Bioinformatics, Department of Biochemistry and Molecular Biology, Indiana University School of Medicine, Indianapolis, IN 46202, U.S.A.;

[2]Molecular Kinetics, Inc., Indianapolis, IN 46268, U.S.A.;

[3]Institute for Biological Instrumentation, Russian Academy of Sciences, 142290 Pushchino, Moscow Region, Russia

## Abstract

This chapter describes a family of intrinsically disordered proteins, members of which fail to form rigid 3-D structures under physiological conditions, either along their entire lengths or only in localized regions. Instead, these intriguing proteins/regions exist as dynamic ensembles where there are significant temporal variations in atom positions and backbone Ramachandran angles. Many of these intrinsically disordered proteins still carry out important biological functions originating from regions that lack specific 3-D structure. The existence of such proteins does not fit the structure-function paradigm, which dominates modern protein science and states that unique 3-D structure is a prerequisite to function. Thus, the classical structure-function paradigm has to be reconsidered to include intrinsically disordered proteins and alternative relationships between protein sequence, structure, and function. This shift in the paradigm represents a real breakthrough for biochemistry, biophysics and molecular biology, as it opens new levels in understanding of the complex life of proteins. This chapter will try to answer the following questions: How intrinsically disordered proteins were discovered? Why these proteins do not fold? What is so special about intrinsic disorder? What are the functional advantages of disordered proteins/regions? What is the functional repertoire of these proteins? What is the relationship between intrinsically disordered proteins and human diseases?

[*] E-mail address: vuversky@iupui.edu; Phone: 317-278-9194; Fax: 317-274-4686; Department of Biochemistry and Molecular Biology, Indiana University School of Medicine, 635 Barnhill Drive, MS#4021, Indianapolis, IN 46202, USA (Corresponding author)

# Introduction: How Intrinsically Disordered Proteins were Discovered

Proteins are the major components of the living cell, which play crucial role in the maintenance of life, and dysfunction of which may cause development of different pathological conditions. Proteins possess endless variety of biological functions but one class of them, known as enzymes, has attracted the major attention of researchers at early days of protein science. This happened mostly because of the fact that enzymes are biological catalysts (i.e., substances which are speeding up a chemical or biochemical reaction which, without the catalyst, would have occurred anyway but at a much slower rate. The catalyst is never used up in the reaction, so that there is always the same amount at the start and the end of the reaction), which catalyze the vast majority of life-related chemical reactions.

For a very long time it was believed that the specific functionality of a given protein is predetermined by its unique 3-D structure. This hypothesis is based on the analogy of a rigid active site with an inorganic catalyst, which is known to accelerate the chemical reaction by providing a lower energy pathway between the reactants and the products. This usually involves the formation of an intermediate, which cannot be formed without the catalyst. The formation of this intermediate and subsequent reaction generally has a much lower activation energy barrier than is required for the direct reaction of reactants to products. Obviously, the active surface of catalyst has to be rigid to do this job efficiently. In other words, the amino acid sequence determines 3-D structure, which, in its turn, determines unique function of a given protein, assuming that following its biosynthesis on the ribosome, a protein must fold to be functional.

The idea that a protein must first fold into 3D structure to carry out function is known as the structure-function paradigm, one of the cornerstones of modern biology, biochemistry, biophysics, molecular biology and medicine, which has been deeply entrenched in protein science well before the resolution of first protein 3-D structure. The primary source of the structure-function paradigm is the "lock and key" hypothesis formulated in 1894 by Emil Fischer to explain the astonishing specificity of the enzymatic hydrolysis of glucoside by different types of similar enzymes, where one enzyme could hydrolyze α- but not β-glycosidic bonds, and another could hydrolyze β- but not α-glycosidic bonds [1]. Based on these observations Fischer [1] wrote (as translated in [2]) "To use a picture, I would like to say that enzyme and glucoside have to fit to each other like a lock and key in order to exert a chemical effect on each other." In this analogy, the lock is the enzyme, the key-hole is the active site of enzyme, and the key is the substrate. Similarly to a situation when only the correctly shaped key opens a particular lock, it has been hypothesized that only the correctly shaped/sized substrate (key) could fit into the key-hole (active site) of the particular lock (enzyme), whereas smaller keys, larger keys, or incorrectly positioned teeth on keys (incorrectly shaped or sized substrates) do not fit into the lock (enzyme).

For sometime, the validity of "lock and key" model and the sequence-structure-function paradigm became an absolute truth after the crystal structures of proteins started to be solved by X-ray diffraction method. In fact, the first determined 3-D structure of an enzyme, lysozyme, where the bound inhibitor was co-crystallized with the protein, immediately showed that the precise locations of certain amino acid side chains is what facilitates catalysis [3]. Since the first reports on X-ray crystallographic structures at atomic resolution for myoglobin [4, 5] and lysozyme [3], more than 29,500 protein structures have been deposited into the Protein Data Bank [6] as of May 31, 2005, with 1,993 structures being deposited in the PDB archive during the first quarter of 2005. These many structures seemed to continue to reinforce a static view of functional protein structure, with the enzyme active site being considered to be a rigid and sturdy lock, providing an exact fit to only one substrate (key). This is because of the fact that the pictures of protein structures that emerge from X-ray crystallography and NMR seem rigid and static. It has been pointed out that the conformational forces, stabilizing the secondary structure and the tertiary fold, are weak and there is enough energy available at ambient temperatures to break some interactions. When these weak interactions are broken, the groups that were involved in these interactions are released and can make new interactions of comparable energy. These rearrangements occur usually in a time scale that it faster than the time required for structure determination by such methods as X-ray crystallography. As a result, 3-D structures of proteins determined by the majority of physical techniques are averaged pictures [7].

Rahter than being highly rigid structures, in reality proteins are highly flexible and dozens of examples of functional disordered regions have been reported. Overall, X-ray and NMR structure determination studies and the characterization of protein structure by other biophysical techniques show that not all proteins are structured throughout their entire lengths. For example, many proteins which have structural information deposited in the Protein Data Bank (PDB) have portions of their sequences missing from the determined structures (missing electron density) [8, 9]. A common reason for missing electron density is that the unobserved atom, side chain, residue, or region fails to scatter X-rays coherently due to variation in position from one protein to the next, e.g. the unobserved atoms are *flexible* or *disordered*. Other proteins happened to represent a big challenge for protein crystallographers because of their flexible and very dynamic nature. Myelin basic protein (MBP) exemplifies these troublemakers [10]. One exhaustive series of attempts to crystallize MBP for X-ray diffraction has been reported, where the authors tried 4,600 different crystallization conditions but were unable to induce crystallization of MBP [11]. Based on these observations it has been concluded that the myelin basic protein falls into the category on "uncrystallizable" proteins. It can be safely assumed that many other unsuccessful crystallization attempts for numerous other proteins have not been reported, since negative results are generally assumed to be unsuitable for publication.

Analysis of protein solution structure by NMR and comparison of the results of the X-ray and NMR analyses of the same protein revealed that solution and crystal structures can be different, with solution one been more flexible [12, 13]. In fact, NMR spectroscopy is able to visualize structure of relatively flexible protein fragments, the electron density of which is missing in crystallographic experiments. Similar conclusions have been made for some proteins analyzed by several biophysical techniques sensitive to the different levels of protein structure. Historically, proteins with flexible structure have been discovered one-by-one and thus each was considered as a unique exception to the "lock and key" rule. They were observed as unique cases of polyfunctional proteins (e.g., serum albumin [14]), or polypeptides with unusual amino acid compositions (e.g., prothymosine $\alpha$ [15-17]), or proteins involved in the binding of large partners (RNA, DNA, proteins, and heme, e.g., histones [18], ribosomal proteins [19], myoglobin [20] and cytochrome *c* [21, 22]) or large quantities of small ones (e.g., osteocalcin [23]). Furthermore, it was suggested that this increased conformational flexibility can be of functional significance. In other words, these data indicated that sometimes proteins do not need to be rigid to be functional. Although these observations were in obvious contradiction with the protein structure-function paradigm, the tipping point for a concept change did not come yet and the importance of disorder was underestimated. In part, this was due the strong prevalence of the lock and key paradigm and the relatively small number of contrary examples, as the information about "non-traditional" proteins was scattered in literature and the vast majority of them have been discovered one-by-one. Thus, for a very long time each protein with highly unusual structural properties and/or strange conformational behavior was considered as an exception from the general rule stating that the function requires rigid 3-D structure.

The importance of flexible structure for some functions also emerged from studies on protein folding. In fact, it has been pointed out that partially structured folding intermediates (such as molten globule and pre-molten globule), which preserve some main elements of native secondary structure and their crude mutual positions in 3-D space, but differ from the rigid globular state by a less tight packing of side chains and by the dramatic increase in the mobility of loops and ends of chain, are almost ideal for some protein functions. Based on these observations it has been suggested that molten globule can exist in a living cell and can be involved in a number of physiological processes [24-26]. The validity of this assumption has been confirmed experimentally, emphasizing the involvement of partially folded intermediates in such processes as interaction with chaperones [27], protein insertion into membranes [28, 29] and interaction with ligands (summarized in [30, 31]). It is necessary to note, however, that even when the functionality has been attributed to the molten globule- or pre-molten globule-like conformations, the major emphasis still was on a concept of rigid 3-D structure. It has been hypothesized that the functional partially folded intermediates in a cell represent kinetic folding intermediates trapped by chaperones just after the protein biosynthesis before proteins can completely fold [24-26], or appear as a result of point mutations preventing polypeptides from complete folding [26, 32]. Some other proteins (such as pore-forming domains of some toxins, or proteins that acts as carriers of large hydrophobic ligands) were assumed to have originally a rigid structure but forced somehow to denature to fulfill their functions [25, 26].

The situation has changed recently mostly due to the efforts of four research groups, which almost simultaneously and completely independently came to the important conclusion that naturally flexible proteins, instead of being just rare exceptions, represent a new and very broad class of proteins [33-36]. Interestingly, this important conclusion was reached commencing from absolutely different starting points and was based on rather different areas of expertise: bioinformatics (Dr. A.K. Dunker's group), NMR spectroscopy (Dr. P.E. Wright's group), protein folding/misfolding (Dr. V.N. Uversky's group) and protein structural characterization (Dr. P. Tompa's group). Many proteins have been shown to lack rigid 3-D structure under physiological conditions *in vitro*, existing instead as dynamic ensembles of interconverting structures. These naturally flexible proteins are currently known as intrinsically disordered [35], natively denatured [37], natively unfolded [38], intrinsically unstructured [33], and natively disordered [39]). We will call these proteins "intrinsically disordered" (ID) from now on. By "intrinsic disorder" we mean that the protein exists as a structural ensemble, either at the secondary or at the tertiary level. In other words, in contrast to ordered proteins whose 3-D structure is relatively stable and Ramachandran angles vary slightly around their equilibrium positions with occasional cooperative conformational switches, intrinsically disordered proteins or regions exist as dynamic ensembles in which the atom positions and backbone Ramachandran angles vary significantly over time with no specific equilibrium values and typically undergo non-cooperative conformational changes. Both extended (random coil-like) regions with perhaps some secondary structure and collapsed (partially folded or molten globule-like) domains with poorly packed side chains are included in our view of intrinsic disorder [35].

Since publication of key studies and reviews describing this new concept [33-36], the literature on these proteins is virtually exploding (see Figure 1). Bioinformatics studies indicated that about 25 to 30% of eukaryotic proteins are mostly disordered [40], that more than half of eukaryotic proteins have long regions of disorder [40, 41], and that more than 70% of signaling proteins have long disordered regions [42].

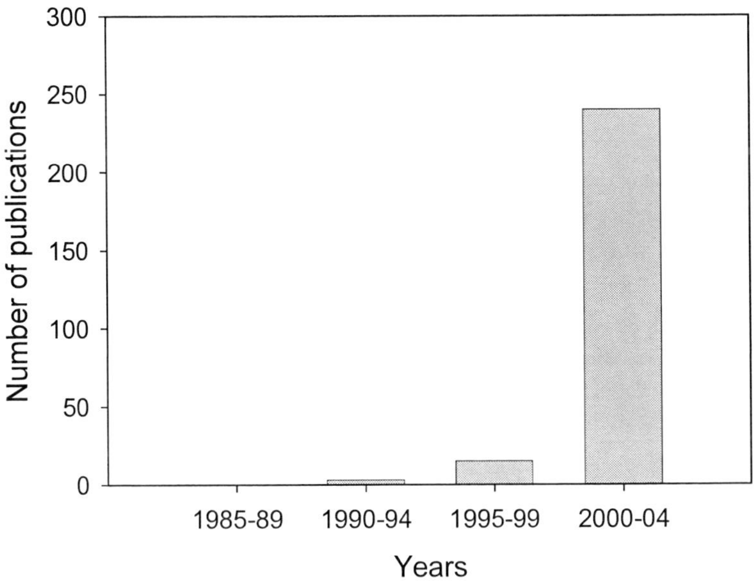


Figure 1. Time-dependent increase in the number of PubMed hits dealing with ID proteins. The following keywords have been used to perform this search: intrinsically disordered, natively unfolded, intrinsically unstructured, intrinsically unfolded and intrinsically flexible.

Now it is recognized that ID is a very abundant phenomenon. In fact, many proteins were shown to contain regions of disorder or even to be entirely disordered. Uversky et al. compiled a list of 91 disordered proteins characterized by NMR, circular dichroism or other biophysical techniques [34]. A subsequent searches of X-ray crystal structures and the literature have expanded this list to more than 200 proteins that contain disordered regions of 30 consecutive residues or longer as characterized by X-ray crystallography, proteolytic digestion or other physical analyses such as NMR or circular dichroism [35]. The commonness of intrinsic disorder was estimated by predicting disorder for whole genomes, including both known and putative protein sequences (see below for the discussion of the disorder predictors). Such predictions have been published for 31 genomes that span the 3 kingdoms. The percentage of sequences in each genome with segments predicted to have $\geq 40$ consecutive disordered residues was used to gain an overview of proteomic disorder. For so many consecutive predictions of disorder, the false positive error rate was estimated from ordered proteins to be less than 0.5% of the segments of 40 and less than 6% of the fully ordered proteins [35, 41]. The eukaryotes exhibited more disorder by this measures than either the prokaryotes or the archaea, with *C. elegans*; *A. thaliana*; *S. cerevisiae*; and *D. melanogaster* predicted to have 52-67% of their proteins with such long predicted regions of disorder, while bacteria and archaea were predicted to have 16-45% and 26-51% of their proteins with such long disorder regions, respectively [40, 41]. The increased amount of disorder in the eukaryota may be related to the increase in cellular signaling in the eukaryota [40, 41, 43]. The functional repertoire and advantages of intrinsic disorder will be discussed below.

## Why ID Proteins Do not Fold

Similar to the "normal" protein for which it has been shown that the correct folding into the rigid biologically active conformation is determined by amino acid sequence, the absence of rigid structure in the "non-traditional" ID proteins may also be somehow encoded in the specific features of their amino acid sequences. In fact, some of the ID proteins have been discovered due their unusual amino acid sequence compositions and the absence of regular structure in these proteins has been explained by the specific features of their amino acid sequences including the presence of numerous uncompensated charged groups (often negative); i.e., a large net charge at neutral pH, arising from the extreme pI values in such proteins [16, 38, 44], and a low content of hydrophobic amino acid residues [16, 44]. Interestingly, the first predictor of intrinsic disorder was developed by R. J. P. Williams based on the abnormally high charge/hydrophobic ratio for the two ID proteins; i.e., using the same set of attributes, large net charge and low overall hydrophobicity [45]. Although this predictor was used to separate just two ID proteins from a small set of ordered proteins, this paper is significant as being the first indication that ID proteins have amino acid compositions that differ substantially from those of proteins with 3-D structure.

Later, this approach was re-invented in a form of charge-hydropathy plot [34]. To this end, 275 natively folded and 91 natively unfolded proteins (i.e., proteins which at physiologic conditions have been reported to have the NMR chemical shifts of a random-coil, and/or lack significant ordered secondary structure (as determined by CD or FTIR), and/or show hydrodynamic dimensions close to those typical of an unfolded polypeptide chain) have been assembled from the literature searches. From the comparison of these datasets it has been concluded that the combination of low mean hydrophobicity and relatively high net charge represents an important prerequisite for the absence of compact structure in proteins under physiological conditions. This observation was used to develop a charge-hydropathy (CH) plot method of analysis that distinguishes ordered and disordered proteins based only on their net charges and hydropathy [34]. Figure 2A represents the original CH-plot and shows that natively unfolded proteins are specifically localized within a unique region of CH phase space. Furthermore, ID and ordered proteins can be separated by a linear boundary, above which a polypeptide chain with a given mean net charge will most probably be unfolded [34]. From the physical viewpoint, such a combination of low hydrophobicity with high net charge as a prerequisite for intrinsic unfoldedness makes perfect sense: high net charge leads to charge-charge repulsion, and low hydrophobicity means less driving force for protein compaction. In other words, these features are characteristic for ID proteins with the coil-like (or close to coil-like) structures. Obviously, such highly disordered proteins represent only a small subset of the ID protein realm.

More detailed analysis was elaborated to gain additional information on the compositional difference between ordered and ID proteins. Comparison of a non-redundant set of ordered proteins with several datasets of disorder (where proteins were grouped based on different techniques, X-ray crystallography, NMR and CD, used to identify disorder) revealed that disordered regions share at least some common sequence features over many proteins [46, 47]. These differences in amino acid compositions are visualized in Figure 2B. Here, the relative content of each amino acid in a given disordered dataset has been expressed as (Disordered–Ordered)/(Ordered). Thus, negative peaks correspond the amino acids in which the disordered segments are depleted compared with the ordered ones, and positive peaks indicate the amino acids in which ID regions are enriched [35]. The arrangement of the amino acids from least to most flexible was based on the scale established by Vihinen et al. [48]. This scale was defined by the average residue B-factors of the backbone atoms for 92 unrelated proteins. Figure 2B shows that the disordered proteins are significantly depleted in bulky hydrophobic (Ile, Leu, and Val) and aromatic amino acid residues (Trp, Tyr, and Phe), which would normally form the hydrophobic core of a folded globular protein, and also possess low content of Cys and Asn residues. The depletion of ID protein in Cys is also crucial, as this amino acid residue is know to have a significant contribution to the protein conformation stability via the disulfide bond formation or being involved in coordination of different prosthetic groups. These depleted residues, Trp, Tyr, Phe, Ile, Leu, Val, Cys and Asn were proposed to be called order-promoting amino acids. On the other hand, ID proteins were shown to be substantially enriched in polar, disorder-promoting, amino acids: Ala, Arg, Gly, Gln, Ser, Glu, and Lys and also in the hydrophobic, but structure-breacking Pro [35, 49, 50]. Note that these biases in the amino acid compositions of ID proteins are also consistent with the low overall hydrophobicity and high net charge characteristic of the natively unfolded proteins (see above).

In addition to amino-acid composition, the disordered segments have also been compared with the ordered ones by various attributes such as hydropathy, net charge, flexibility index, helix propensities, strand propensities, and compositions for groups of amino acids such as W + Y + F (aromaticity). As a result, 265 property-based attribute scales [49] and more than 6,000 composition-based attributes (e.g., all possible combinations having one to four amino acids in the group) have been compared [51]. It has been established that ten of these attributes, including 14 Å contact number, hydropathy, flexibility, β-sheet propensity, coordination number, R+E+S+P, bulkiness, C+F+Y+W, volume, and net charge, provide fairly good discrimination between order and disorder [35]. Thus, the ID proteins differ dramatically from the ordered proteins in their amino acid sequences. These differences were used to develop different predictors of intrinsic disorder.

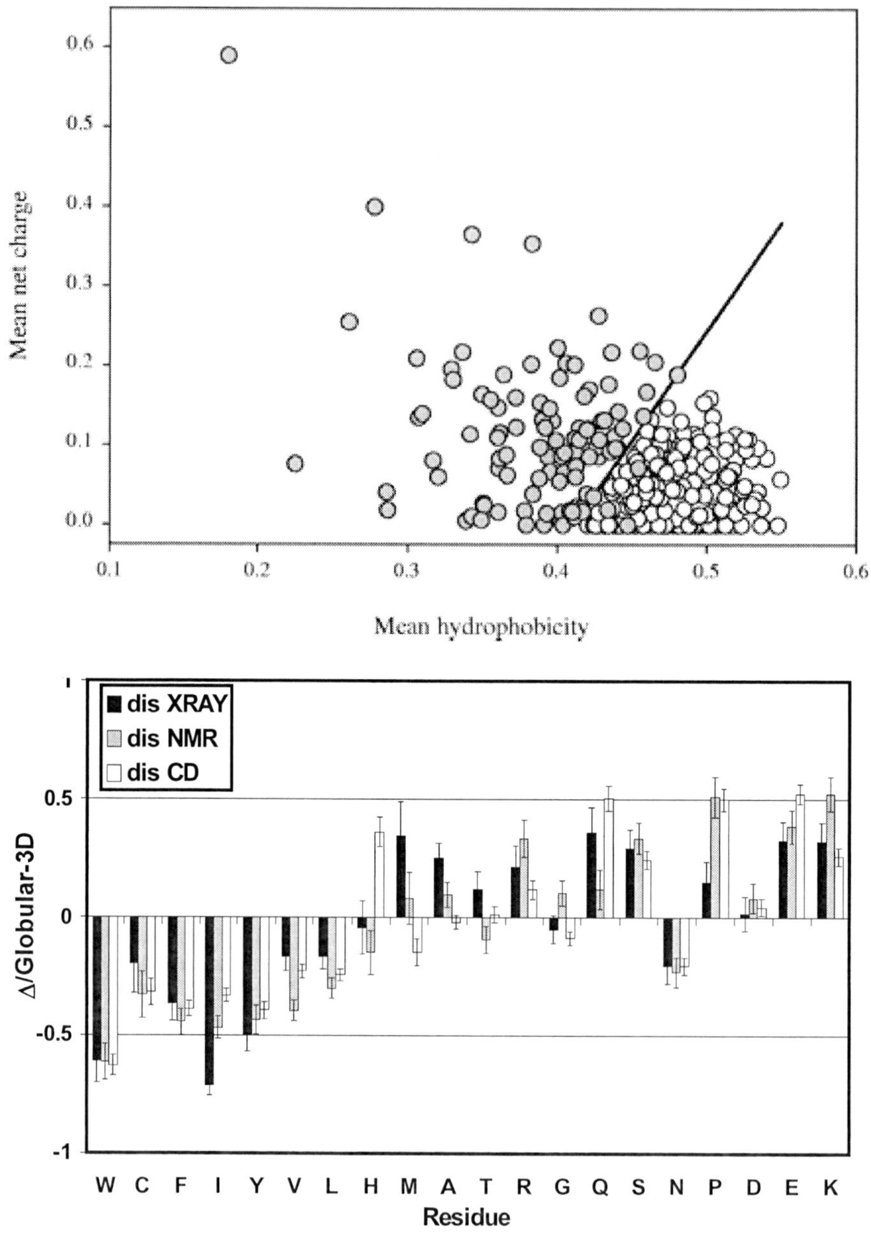


Figure 2. Peculiarities of amino acid composition of ID proteins. A. Comparison of the mean net charge and the mean hydrophobicity for a set of 275 folded (open circles) and 91 natively unfolded proteins (gray circles). The solid line represents the border between intrinsically unstructured and native proteins (see text). **B.** Order/Disorder composition profile. Comparisons of amino acid compositions of ordered protein with each of three databases of disordered protein. The ordinates are (% amino acid in disordered dataset - % amino acid in ordered dataset) / (% amino acid in ordered dataset) = Δ/globular_3D. The residues are ordered according to the Vihinen's flexibility scale [48]. Names of each database indicate how the disordered regions were identified. Negative values indicate that the disordered database has less than order, positive indicates more than order.

## How to Predict ID Protein from Amino Acid Sequence

The fact that the ID proteins differ dramatically from the ordered proteins in their amino acid sequences was used to develop different predictors of intrinsic disorder. In 1997, a first Predictor Of Natural Disordered Regions (PONDR®) that used 10 of the above-mentioned sequence attributes was developed [52]. PONDR® operates from primary sequence data alone, using the nonlinear models (feed-forward neural networks) as the basis for the order/disorder discrimination [52]. While working on PONDR® development, datasets of ordered and disordered protein sequence segments as characterized by X-ray diffraction [52], by NMR [47, 53], by X-ray diffraction partitioned by location [54], or by homology [53, 55] were assembled. Balanced datasets were used to train neural network predictors with various inputs. Predictions of order/disorder on out-of-sample, balanced datasets (e.g. using 5-cross validation), produced accuracies in the range of 70-84% (Table 1). Accuracies on individual proteins can vary by 10% from the averaged values. Recently, our prediction accuracy has been considerably increased using the greatly expanded database of disordered proteins and improved computational techniques [56, 57]. The relatively high prediction accuracies strongly support our use of amino acid sequence to predict disorder as an element of native protein structure. Since the time of its first introduction, PONDR® underwent dramatic development and several versions of the predictors with the increased accuracy and reliability are currently available. This includes PONDR® VL-XT, VL3, and VSL1 predictors, access to which is provided by Molecular Kinetics, Inc. (at http://www.pondr.com/).

Table 1 shows that to achieve balanced accuracies on order and disorder, we were able to improve disorder only slightly while losing accuracy on order (e.g., compare data for VSL1 and VSL2). This is quite interesting, and we think that this effect arises from order-like regions in the disordered parts, so we start losing order accuracy when we start calling these types of sequences to be disordered. Table 1 also shows that we continue to observe better prediction of ordered residues as compared to disordered residues. Our current view on this asymmetry is that regions structurally characterized as disordered often contain local regions with a strong tendency to become ordered – and these do become ordered when the correct binding partner comes along. Many examples of this type of behavior are discussed below. In fact, we noticed some time ago that predictions of order in regions characterized to be disordered often identify binding sites [58]. We believe that this behavior identifies protein structural element or feature that mediates the binding events of initially disordered regions. This element consists of a short region that undergoes coupled binding and folding within a longer region of disorder. We call these features "molecular recognition elements" (MoREs). Examples of MoREs bound to their partners can be found in the α-helix, β-strand, polyproline II helix, or irregular secondary structure conformations, and in various mixtures of the four structural forms. Based on these observations an algorithm has been recently elaborated [59] that identifies regions having a propensity for α-helix-forming molecular recognition elements (α-MoREs) based on a discriminant function that indicates such regions while giving a low false-positive error rate on a collection of structured proteins (see below for more details).

**Table 1. Accuracy and improvement of neural network predictors of natural disordered regions (PONDRs®)**

| Name | Training Set | # Disordered Residues | Accuracies % | |
|---|---|---|---|---|
| | | | Order[a] | Disorder[b] |
| XL1 | 7 X-ray | 502 | 71 | 47 |
| VL1 | 7 NMR, 8 X-ray | 1,366 | 83 | 45 |
| XL-XT | VL1 plus XT[c] | | 71 | 59 |
| VL2 | 53 X-ray,<br>35 NMR,<br>52 CD | 17,978 | 76 | 65 |
| VL3[d] | 54 X-ray,<br>40 NMR,<br>58 CD | 22,434 | 84 | 59 |
| VSL1[e] | 230 long DR[f]<br>983 short DR<br>Ordered regions | 25,958<br>9,632<br>354,169 | 83 | 79 |
| VSL2[g] | 230 long DR<br>983 short DR<br>Ordered regions | 25,958<br>9,632<br>354,169 | 81 | 82 |

[a] O_PDB_S25

[b] Combined dis_X-ray, dis_NMR and dis_CD

[c] XT is a joint name for the N-terminus (XN), and the C-terminus (XC) predictors, which were trained using x-ray crystallographic data, where the terminal disordered regions were 5 or more amino acids in length.

[d] Besides the addition of a few more chains, substantial cleaning of the training databases was carried out between VL2 and VL3. Several incorrectly labeled chains were identified and fixed and order/disorder boundaries were adjusted in a few other proteins

[e] The VSL1 predictor combines two predictors optimized for long (>30 residues) and short (≤30 residues) disordered regions, respectively, using weights generated by a third meta-predictor. The attributes used include amino acid frequencies, sequence complexity, ratio of net charge / hydrophobicity, averaged flexibility, and averaged PSI-BLAST profiles calculated over symmetric input windows.

[f] Disordered region

[g] VSL2 is a slightly improved version of VSL1 predictor. The training data for VSL2 were slightly different: 8 ambiguous sequences were removed; His-tags were not used in training, short DR of 1-3 residues were not used in training. Also, linear SVM instead of logistic regression was used for VSL2 version (Kang Peng, personal communication).

**Note:** Both VSL1 and VSL2 take advantage of length dependencies

Recently, a number of groups have published predictors of protein disorder, several of which are available on the web (see Table 2). Furthermore, as a recognition of the increased interest to this phenomenon, disorder prediction has been included as part of the 5[th] and 6[th] Critical Assessment of Methods for Protein Structure Prediction (called CASP5 and CASP6), with 4 and 20 groups participating in the prediction of disordered regions in CASP5 and CASP6, respectively. It has been emphasized that the ability to predict disorder from sequence with good accuracy in a truly blind experiment adds confidence in the various results obtained by disorder predictions [60]. Here we will discuss just one of them developed by Rost and co-workers [61]. This study was initiated by the observation of long, irregular sequences, or "loopy regions," that are visible in x-ray crystal and NMR structures. Since such regions are observed in X-ray structures, they don't fit our definition of intrinsic disorder. However, such loopy regions typically have no internally buried regions and thus probably exist as disordered regions that become ordered by either crystal contacts or by interactions with the surfaces of globular protein domains. The observed loopy regions were not used as a training set, however, but rather to motivate the development of the algorithm. The resulting Rost algorithm was based upon the absence of predicted secondary structure and the presence of predicted solvent accessibility in fairly long regions. This algorithm is used to indicate the presence of such "loopy" proteins, which were also called regions of "no regular secondary structure" (NORS). The NORS indicator as developed gave a low false positive prediction rate on globular, ordered protein, but its performance has not been evaluated on a set of proteins known to be disordered. Comparisons between prediction results for the NORS algorithm and results for PONDR indicate that these predictors, in fact, overlap significantly, but are by no means coincident (unpublished observations).

**Table 2. Protein disorder predictors**

| Predictor | Web address | Reference |
|---|---|---|
| Charge-hydropathy plot | http://www.pondr.com/ | [34] |
| DisEMBL™ | http://dis.embl.de | [174] |
| DISOPRED | http://bioinf.cs.ucl.ac.uk/disopred/ | [175] |
| DISpro | http://www.ics.uci.edu/~baldig/diso.html | |
| DRIPpred | http://sbcweb.pdc.kth.se/cgi-bin/maccallr/disorder/submit.pl | http://www.forcasp.org/paper2127.html |
| FoldIndex© | http://bioportal.weizmann.ac.il/fldbin/findex | [176] |
| GlobPlot | http://globplot.embl.de/ | [177] |
| IUPred | http://iupred.enzim.hu/ | [178] |
| NORSp | http://cubic.bioc.columbia.edu/services/NORSp/ | [179] |
| PONDR®[a] | http://www.pondr.com/ | [35, 46, 50, 52, 55, 180] |
| PreLink | http://genomics.eu.org/ | [181] |
| RONN Disorder Prediction | http://www.strubi.ox.ac.uk/RONN | [182] |
| SEG[b] | http://mendel.imp.univie.ac.at/METHODS/seg.server.html/ | [183] |

[a] PONDR® is a family of ID predictors, which includes VL-XT, VL3,

[b] Formally, SEG is not a disorder predictor. It is an indicator of low sequence complexity regions

All of the predictors developed so far use as input the protein amino acid sequence and its attributes. The attributes are allied in different combinations and applied to classify each residue within a local sequence region (or in some cases to classify the entire protein) as either ordered or disordered. Different approaches are used to weight and combine the various features. Predictors of disorder are based on various computational approaches, including, analytical algebraic functions, linear least squares, logistic regression, neural networks, and support vector machines. Finally, it is necessary to emphasize that even although modern predictors of intrinsic disorder use different (and in some cases very sophisticated) computational approaches, they are mostly based on the concepts elaborated in the mentioned pioneering studies [34, 35, 49, 52] or some derivatives of these concepts.

It is important to remember that comparing several predictors on an individual protein of interest or on a protein dataset can provide additional insight regarding the predicted disorder if any exists. This is illustrated by our recent study, were two distinct methods for using amino acid sequences to predict which proteins are likely to be mostly disordered, cumulative distribution function (CDF) analysis and charge-hydropathy (CH) plot, have been compared [40]. CDF is based on the PONDR® VL-XT predictor, which predicts the order-disorder class for every residue in a protein [40, 41]. CDF analysis summarizes these per-residue predictions by plotting PONDR® scores against their cumulative frequency, which allows ordered and disordered proteins to be distinguished based on the distribution of prediction scores. The other method of order-disorder classification is charge-hydropathy plots [34], in which ordered and disordered proteins being plotted in charge-hydropathy space can be separated to a significant degree by a linear boundary. It has been established that CDF analysis predicts a much higher frequency of disorder in sequence databases than CH-plot discrimination [40]. However, the vast majority of disordered proteins predicted by charge-hydropathy discrimination were also predicted by CDF analysis. These findings are not a big surprise, as CH-plot analysis discriminates protein using only two attributes, mean net charge and mean hydrophobicity, whereas PONDR® VL-XT (and consequently CDF) is a neural network, which is a nonlinear classifier, trained to distinguish order and disorder based on a relatively large feature space (including average coordination number, amino acid compositions (aromatic and charged residues), and net charge). Thus, CH feature space can be considered as a subset of PONDR® VL-XT feature space [40]. Importantly, these findings may be physically interpretable in terms of different types of disorder, collapsed (molten globule-like) and extended (pre-molten globule- and coil-like). Under this consideration, the CH-plot classification discriminates proteins with the extended disorder from a set of globular conformations (molten globule-like or rigid well-structured proteins) and proteins predicted to be disordered by the CH-plot approach are likely to belong to the extended disorder class. On the other hand, PONDR®-based approaches can discriminate all disordered conformations (coil-like, pre-molten globules and molten globules) from rigid well-folded proteins, suggesting that CH classification is roughly a subset of PONDR® VL-XT, in both predictions of disorder and feature space [40]. Based on this reasoning, several interesting conclusion have been made. It has been suggested that if a protein is predicted to be disordered by both CH and CDF, then, it is likely to be in the extended disorder class. However, a protein predicted to be disordered by CDF but predicted to be ordered by CH-plot might have properties consistent with a dynamic, collapsed chain; i.e., it is likely to be in the native molten globule class. Finally, proteins predicted to be ordered by both algorithms are of course likely to be in the well-structured class [40]. Importantly, the fact that CDF analysis predicts about 2-fold higher frequency of disorder in sequence databases than CH-plot classification suggests that approximately half of disordered proteins in different proteomes possess extended disorder, whereas another half represents proteins with the collapsed disorder [40].

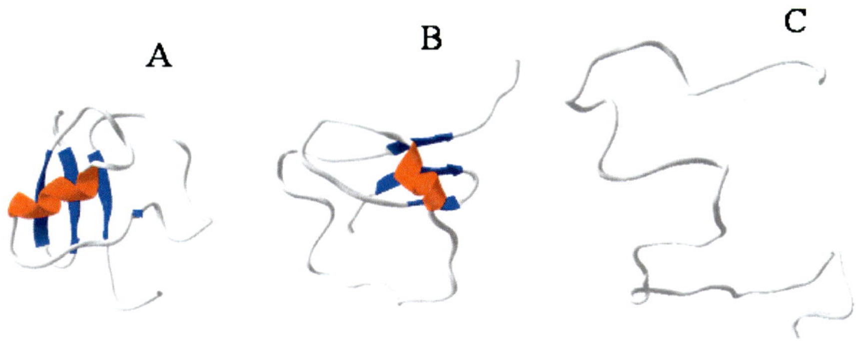


Figure 3. Illustrative examples of ID proteins: A. Collapsed (molten globule-like) disorder; B. Extended (pre-molten globule-like) disorder; C. Extended (coil-like) disorder.

As it has been already mentioned, protein disorder is a multi-faced phenomenon; i.e., disordered proteins, being mobile, flexible, and dynamic, might have very different structural features, which range from collapsed molten globule-like conformation to extended coil-like state. It has been suggested that just as an ordered protein is comprised of different types of secondary structure ($\alpha$-heices, $\beta$-strends, $\beta$-turns, $3_{10}$-helices and others), ID protein can also be made up of distinguishable types of disorder [62]. To check this hypothesis, a novel partitioning algorithm based on the differential prediction accuracies has been developed [62]. This algorithm used the notion that a specialized predictor built on a given disorder flavor should have significantly higher same-flavor accuracy than other-flavor predictors or than a global predictor applied to the same given flavor. Application of this partitioning algorithm to known disordered proteins identified three distinctive flavors of disorder, arbitrarily called V, C, and S [62]. Importantly, the flavor-specific disordered proteins have been shown to be distinguishable not only by their amino acid compositions, but also by disordered sequence locations, and biological functions. Based on these observations, it was proposed that specific flavor-function relationships do exist and thus it is possible (in principle) to identify the functions of disordered regions from their amino acid sequences alone, without any need for specific structural knowledge [62].

## What Is so Special about ID Proteins

ID proteins have dynamic structures that interconvert on a number of timescales and were shown to have many similarities to non-native states of "normal" globular proteins, which may exist in at least four different conformations: native (ordered), molten globule, pre-molten globule, and coil-like [63-66]. Using this analogy, it has been established that ID proteins and regions under physiological conditions *in vitro* might contain collapsed-disorder (i.e., where ID is present in a form of molten globules) and extended-disorder (i.e., regions where ID is present in a form of random coil or pre-molten globule) [35, 39, 66]. This is illustrated by Figure 3, which schematically represents the three types of disorder mentioned above, molten globule, pre-molten globule and coil.

*Collapsed disorder.* The structural properties of the molten globule are well known and have been systematized in number of reviews (e.g., see [67] and references therein). The protein molecule in this intermediate state has no (or has only a trace of) rigid cooperatively melted tertiary structure. However, it is characterized not only by the well-developed secondary structure, but also by the presence of some topology, i.e., relatively fixed mutual positioning of the secondary structure elements [68-75]. A considerable increase in the accessibility of a protein molecule to proteases was noted as a specific property of the molten globule [76-80]. It was also shown that transformation into this intermediate state is accompanied by a considerable increase in the affinity of a protein molecule to the hydrophobic fluorescence probes (such as 8-anilinonaphthalene-1-sulfonate, ANS) and this behavior should be considered as a characteristic property of the molten globule state [81]. It has been established that the averaged value for the increase in the hydrodynamic radius in the molten globule state compared with the native state is no more than 15% which corresponds to volume increase of ~50% [26, 66, 67, 82]. Small-angle X-ray scattering analysis revealed that the protein molecule in this partially folded state has a globular structure typical of ordered globular proteins [83-87].

*Extended disorder.* A significant number of sequences encodes for the extendedly disordered proteins that are characterized by the low sequence complexity. Are these proteins random coils, or do they possess residual structure? If they have residual structure, how should they be classified? Based on the analysis of the available literature, it has been concluded that such proteins do not possess uniform structural properties, as expected for members of a single thermodynamic entity. In fact, they may be divided into two structurally different groups, intrinsic coils and intrinsic pre-molten globules [66, 88, 89]. Proteins from the first group have hydrodynamic dimensions typical of considerably unfolded polypeptide chain in poor solvent (see below), and do not possess any (or almost any) ordered secondary structure. Proteins from the second group are more compact (see below), and exhibit some amount of residual secondary structure. However, they are still less dense than native globular or molten globule proteins [66, 88, 89]. Other properties of extendedly disordered proteins can be derived from structural features described below for unfolded and pre-molten globular conformations of globular proteins.

It is important to remember that unfolded protein is never random coil. In fact, the existence of significant residual structure in the unfolded globular protein has been described even under the most severe denaturing conditions, such as high concentrations of strong denaturants [90-93]. Thus, coil-like ID proteins are not completely random, but are characterized by the presence of some residual (and highly flexible) structure. This fact is very important for the functioning of these proteins (see below).

The structural peculiarities of a polypeptide chain in the pre-molten globule state are briefly outlined below. The protein molecule in this state has no rigid tertiary structure. It is characterized by a considerable secondary structure, although much less pronounced than that of the native or the molten globule protein (protein in the pre-molten globule state has ~50% native secondary structure, whereas in the molten globule state the corresponding value is close to 100%). The protein molecule in the pre-molten globule state is considerably less compact than in the molten globule or native states, but it is still more compact than the random coil (its hydrodynamic volume in the molten globule, the pre-molten globule and the unfolded states, in comparison that of the native state, increases 1.5, ~3 and ~12 times, respectively). The protein molecule in the pre-molten globule state can effectively interact with the hydrophobic fluorescent probe ANS, though essentially weaker than in the molten globule state. This means that at least part of the hydrophobic clusters of polypeptide chain accessible to the solvent is already formed in the pre-molten globule state [26, 63, 65, 67, 94]. It has been also established that in the pre-molten globule state the protein molecule has no globular structure [86, 87, 94]. The last observation indicates that the pre-molten globule probably represents a "squeezed" and partially ordered form of the coil [66, 87, 95, 96]. Importantly, several structural elements of these squeezed coils may occupy native-like positions [63, 65, 87]. This fact is of functional importance too.

*Conformational behavior.* Because of the ID proteins possess strong biases in their amino acid compositions one might expect that this will be reflected in their conformational behavior. It has been established that an increase in temperature induces the partial folding of intrinsically unstructured proteins (i.e., proteins with extended disorder), rather than the unfolding typical of globular proteins. The effects of elevated temperature may be attributed to increased strength of the hydrophobic interaction at higher temperatures, leading to a stronger hydrophobic driving force for folding (summarized in [89]). The response of proteins and protein fragments with collapsed disorder on temperature increase is different. Sometimes heating does not affect these proteins at all and they preserve their disordered conformation in a wide temperature range. In some cases melting can be observed. However, this process is much less cooperative than melting of ordered globular proteins.

For a number of extendedly-disordered proteins it has been shown that a decrease (or increase) in pH induces partial folding of intrinsically unordered proteins due to the minimization of their large net charge present at neutral pH, thereby decreasing charge/charge intramolecular repulsion and permitting hydrophobic-driven collapse to the partially-folded conformation (summarized in [89]).

Importantly, this high temperature and extreme pH stability of ID proteins can be used isolate them from cell extracts. For example, disordered proteins with extended disorder can be separated from ordered proteins by their intrinsic indifference to denaturing conditions that originates from the lack of tertiary and secondary structure. In other words, ID proteins can be isolated from ordered ones as proteins which "survived" (i.e., remained soluble) harsh denaturing conditions usually leading to the precipitation of ordered proteins, including extensive heating/boiling or incubation in the presence of trichloroacetic acid (TCA), or perchloric acid (PCA) [97].

Finally, unique electrophoretic mobility of ID proteins should be mentioned. Electrophoresis is mostly applied either to determine the molecular mass of proteins or to elucidate the charge difference and/or form of the macromolecule. However, it has been pointed out that due to their unique amino acid compositions, ID proteins bind less sodium dodecyl sulphate (SDS) than "normal" proteins [36]. As a result, they possess abnormal mobility in SDS polyacrylamide gel electrophoresis experiments and their apparent molecular masses determined by this technique are often 1.2-1.8 times higher than real one calculated from sequence data or measured by mass spectrometry (for example, see [98].

**Table 3. Functional classes of disorder. Based on data reported in [36]**

| Class | Example | Function |
|---|---|---|
| **Entropic chains** | Microtubule-associated protein 2 | Entropic bristle, spacing in microtubule architecture |
| **Effectors** | 4EBP1, 2, 3 | Inhibitor of translation initiation |
| **Scavengers** | Caseins | Inhibition of calcium precipitation in milk |
| **Assemblers** | Caldesmon | Actin polymerization |
| **Display sites** | CREB transactivator domain | Regulation by phosphorylation |

# What is the Functional Repertoire of ID Proteins

As it has been already discussed, intrinsic disorder is very common in nature. Figure 4 further illustrates this conclusion showing that intrinsic disorder is abundant at the proteome level, and many proteins were not only predicted to have long disordered regions, but were prognosticated to be disordered in whole. These observations suggest that although intrinsically disordered proteins fail to form fixed 3D-structures under physiological conditions, they might carry out important biological functions, the fact which was recently confirmed by several comprehensive studies [33-36, 42, 81, 88, 89, 99-104]. Furthermore, sites of posttranslational modifications (acetylation, hydroxylation, ubiquitination, methylation, phosphorylation, etc.) and proteolytic attack are frequently associated with regions of intrinsic disorder [35].

In agreement with this hypothesis, more than 150 proteins have been recently identified as containing functional disordered regions, or being completely disordered, yet performing vital cellular roles [35, 105]. Twenty-eight separate functions were assigned for these disordered regions, including molecular recognition via binding to other proteins, or to nucleic acids [99, 105]. An alternative view is that functional disorder fits into at least five broad classes based on their mode of action [36]. These broad classes, with illustrative examples, are outlined in Table 3. Recent research added chaperones to this list [36].

Obviously, for these proteins, the predominant structure-function paradigm is insufficient, which suggests a more comprehensive view of the protein structure/function relationships is needed. In fact, a new paradigm was recently offered [35, 43, 99] to elaborate the sequence-to-structure-to-function scheme in a way that includes the novel functions of disordered proteins (Figure 5). The complex data supporting this revised view were summarized in The Protein Trinity Hypothesis [43], which suggested that native proteins can be in one of three states, the solid-like ordered state, the liquid-like collapsed-disordered state or the gas-like extended-disordered state. Function is then viewed to arise from any one of the three states or from transitions between them. This model was subsequently expanded to include a fourth state (pre-molten globule) and transitions between all four states [88].

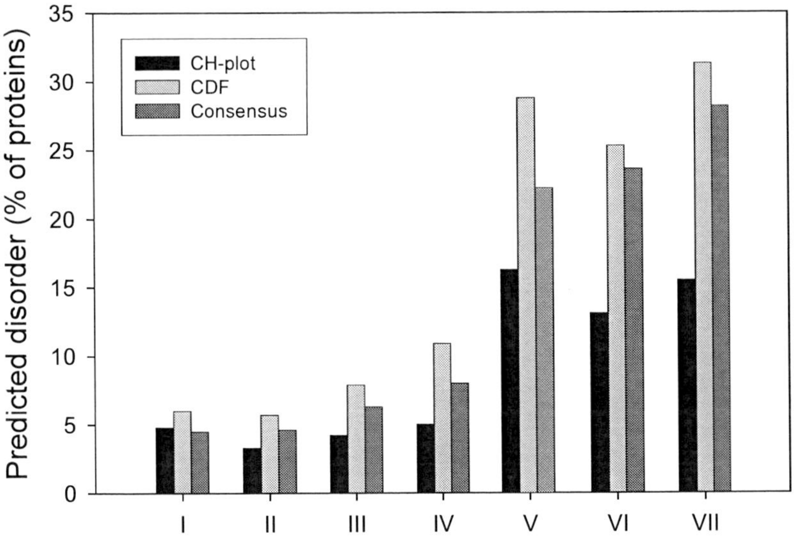


Figure 4. Predicted abundance of mostly disordered proteins in several proteomes: I, *Y. pestis*; II, *E. coli*; III, *A. fulgidus*; IV, *M. thermoautotrophicum*; V, *S. cerevisiae*; VI, *A. thaliana*; VII, *M. musculus*. Analysis was performed by three predictors of mostly disordered proteins: the charge-hydropathy (CH) plot, the cumulative distribution function (CDF) of PONDR® VL-XT score, and a consensus predictor that combines the CH-plot and CDF predictors. The main point is that eukaryotes appear to contain far more intrinsic disorder as compared to prokaryotes. This amount of predicted disorder has important functional consequences, and so proteomic experiments need to be redesigned to recognize and explore intrinsically disordered proteins.

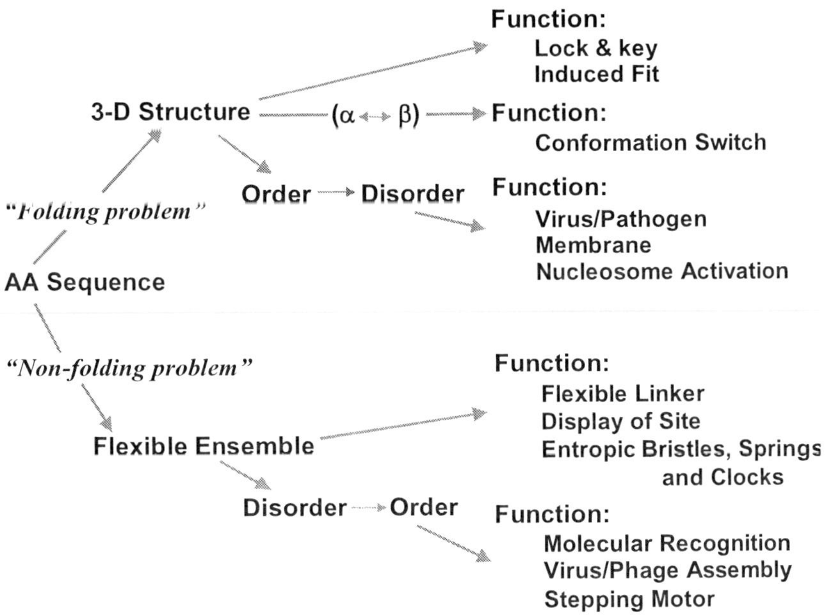


Figure 5. Involvement of intrinsic disorder in protein function. Note that the classical structure-function paradigm cannot describe many of the function proteins perform.

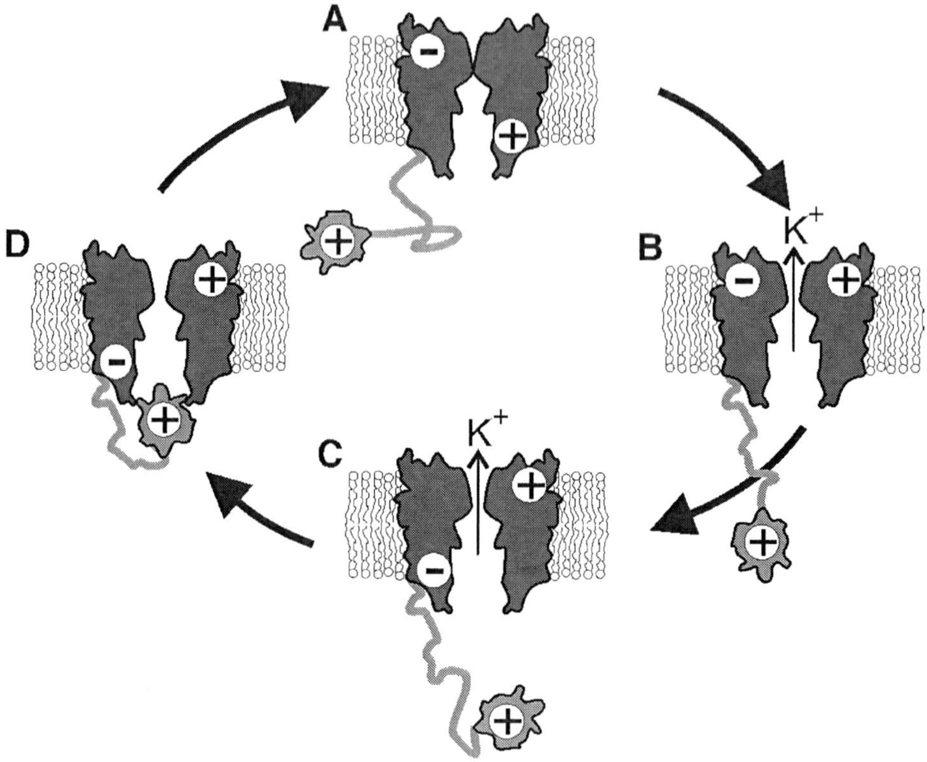


Figure 6. Example of an entropic clock: Simplified model of a Shaker-type voltage-gated $K^+$ ion channel (blue) with 'ball and chain' timing mechanism. The 'ball and chain' is comprised of an inactivation, or ball, domain (yellow) that is tethered to the pore assembly by a disordered chain (red) of ~ 60 residues. For simplicity, only four of the proposed ten states are shown. The cytoplasmic side of the assembly is oriented downward. **A.** Closed state prior to membrane depolarization. Note that conformational changes of the pore have sealed the channel and a positive charge on the cytoplasmic side of the pore assembly excludes binding of the ball domain. **B.** Open state following membrane depolarization. **C.** After depolarization, the cytoplasmic side of the pore opening assumes a negative charge that facilitates interaction with the positively charged ball domain. **D.** Inactivation of the channel occurs when the ball domain occludes the pore. The transition from C to D does not involve charge migration and can be modeled as a random walk of the ball domain towards the pore opening. (Portions of figure based on Antz et al. [106]).

It is necessary to note that some important activities related to molecular recognition do not directly involve coupled binding and folding, but rather are dependent on the flexibility, pliability and plasticity of the backbone. We are calling these "entropic chain activities", as they rely entirely on an extended random-coil conformation of a polypeptide that has to maintain constant motion during functioning. Discovering new entropic chain activities and estimating the fraction of ID regions involved in such activities are both intriguing problems [35]. One illustrated example of such entropic chain activities is considered below. Voltage gated ion channels cycle among three states: closed (sensitive to voltage), open, and inactive (insensitive to voltage). In the ball and chain mechanism for ion channel inactivation, a highly flexible "chain" carries out a random search until the "ball" plugs the open channel (Figure 6). Portions of this figure are based on findings of Antz et al. [106]. For simplicity, only four of the proposed ten states are shown [107]. The inactivation depends on a binding interaction between the channel opening and the "ball". The time of opening is also a crucially important. This "time of being open" function depends directly on the length and flexibility of the disordered "chain." An extended disordered region functions as one component of an entropic clock found in some ion channels. Charge migrations within the tetrameric pore proteins are associated with the majority of state changes of voltage-gated $K^+$ ion channels [107]. However, the timing of the inactivation step is determined by the time it takes for a mobile domain to find and block the channel. The movement of the mobile domain is restricted by a tether composed of ~ 60 disordered residues (Figure 6). The timing of channel inactivation is a function of the length of the disordered tether. Since ion channels serve to modulate the excitability of nerve cells, their malfunction can have substantial impact on human health. Mutations in the human homolog of the Shaker K+ channel (KCNA1) can lead to myokymia, partial epilepsy, or episodic ataxia [108, 109]. The ball and chain model was originally developed from experiments showing that protease digestion caused the open state to remain open, and then adding back the trypsin-released peptide led to channel inactivation. Recent NMR data provides direct confirmation of the flexibility of the "chain" region. If channel closure were to involve a random search by a uniformly flexible chain, the time of closure would be inversely proportional to the square of the length of the chain [110]. Genetic engineering of the Shaker $K^+$ channel was used to construct channels chains of various lengths. The inactivation times for the chains followed the expected dependence on length. Shorter chains speeded up inactivation and longer chains slowed inactivation [111]. However, the predicted dependence on the square of the length of the chain was not followed. Failure to follow the simplified model is not surprising because the flexibility along the chain is not uniform as in the model. Thus, overall these data strongly support the ball and chain mechanism.

It has been pointed out that in a living organism proteins participate in complex interactions, which represent the mechanistic foundation of the organism's physiology and function. Regulation, recognition and cell signaling involve the coordinated actions of many players. To achieve this coordination, each participant must have a valid identification (ID) that is easily recognized by the others. For proteins, these IDs are often within intrinsically disordered regions [103, 104]. Thus, intrinsically disordered regions are typically involved in regulation, recognition, signaling and control pathways in which interactions with multiple partners and high-specificity/low-affinity interactions are often requisite. In this way, the functional diversity provided by disordered regions complements those of ordered protein regions.

*ID proteins as hubs in protein signaling networks.* It is known that many diverse systems may best be described as networks with complex topologies, which is often assumed to be either completely regular or completely random [112]. An important feature of the random networks (such as the highway system) is that, despite the random placement of links, most nodes have approximately same number of links in the resulting systems [112, 113]. In fact, in a random network the nodes follow a Poisson distribution with a bell shape. Nodes that have significantly more or fewer links than the average are difficult to find in such a system [113]. On the other hand, certain other networks, known as "scale-free" or "small-world" networks, are different: they have hubs, with many connections, and ends, that aren't connected to anything but a hub (if highways were like this, there would be a lot of dead-ends) [114, 115]. Scale-free networks combine the local clustering of connections characteristic of regular networks with occasional long-range connections between clusters, as can be expected to occur in random networks. Thus, some nodes have a tremendous number of connections to other nodes, whereas most nodes have just a few [113]. The popular nodes, called hubs, can have hundreds, thousands or even millions of links. It has been emphasized that from this perspective, the network appears to have no scale [113]. For such scale-free networks the distance between nodes follows a power-law distribution [116]. Examples of scale-free networks include the author-collaboration network, the airline routes, the World Wide Web, the metabolic network, the protein domain networks, and, apparently, the protein signaling networks inside cells (systematized in [115]). It has been recently suggested that α-synuclein, p53 and many other ID proteins that interact with large numbers of distinct partners form hubs in the scale-free protein-protein interaction network inferred for the cell [103, 104]. Recent bioinformatics studies support the importance of disorder for hub proteins [183-186] or for the partners that bind to structured hubs [187,188].

## What are the Functional Advantages of ID Proteins/Regions

Importantly, even sturdy locks (i.e., protein active sites) were shown to be rather flexible. In fact, as early as in 1958 it has been recognized that some compounds, being able to bind to the enzyme, were not processed to give a product. Some enzymes seemed to require smaller or larger degree of flexibility to be functional. To account for these discrepancies, a modification of the "lock and key" model called the "induced fit" theory has been proposed by Koshland [117]. According to this theory and its subsequent modifications/interpretations, the enzyme is partially flexible and the substrate does not simply bind to the active site, but it has to bring about changes to the shape of the active site to activate the enzyme and make the reaction possible. Substantial experimental evidence has been accumulated to support this view. For example, the existence of functional flexibility within the active site has been demonstrated by X-ray crystallographic analysis of *E. coli* dihydrofolate reductase liganded with different cofactors and substrates. In fact, Sawaya and Kraut have analyzed crystal structures of different forms of this protein, including the holo-enzyme, the Michaelis complex, the ternary product complex, the tetrahydrofolate binary complex as well as the tetrahydrofolate--NADPH complex. These structures can be used to reconstruct a 2.1 Å resolution movie, depicting the sequence of events during the catalytic cycle, which showed that the enzyme adopts different conformational states while complexed with different ligands, suggesting that the process of enzymatic catalysis might be accompanied by significant conformational changes [118].

It has been emphasized that signaling and regulation is among the most important function of intrinsically disordered (ID) proteins [42]. Qualitatively, it seems reasonable that highly mobile proteins would provide a better basis for signaling and recognition. For example, disordered regions can bind partners with both high specificity and low affinity [119]. This means that the regulatory interactions can be specific and also can be easily dispersed. Obviously this represents a keystone of signaling – turning a signal off is as important as turning it on [43]. Another crucial property of ID proteins for their function in signaling networks is binding diversity; i.e., their ability to partner with many other proteins and other ligands, such as nucleic acids [120]. This opens a unique possibility for one regulatory region or one regulatory protein to bind to many different partners. In agreement with this hypothesis it has been shown that proteins making multiple interactions are more likely to lead to lethality if deleted [121]. There are several other reasons of why ID proteins might be superior to their ordered counterparts. This includes, but is not limited to: binding commonality in which multiple, distinct sequences recognize a common binding site (with perhaps different folds in the various complexed ID proteins) [122]; the ability to form large interaction surfaces as the disordered region wraps-up [123] or surrounds its partner [124]; faster rates of association by reducing dependence on orientation factors and by enlarging target sizes [125]; and faster rates of dissociation by unzippering mechanisms [35].

An interesting consequence of the capability of ID proteins to interact with different binding partners is their polymorphism in bound state; i.e., an ID protein (or ID region) might have completely different geometries in the rigidified structures induced by the binding to its partner, depending on the nature of the bound partner. Recent crystallographic studies on glycogen synthase kinase 3β (GSK3β), a Ser/Thr protein kinase and its interactions with FRAT1 and axin provide an illustrative example of these polymorphic bound states [126]. It has been shown that a sharp turn breaks the structure of FRAT peptide into two distinct and separate α-helical segments, whereas the axin peptide is bound as a single unbroken α-helix with a turn and an irregular tail [126]. Overall, despite the fact that the primary binding sites for axin and FRAT on GSK3β have been found to overlap substantially in the crystal structures, so that their binding is mutually exclusive, the GSK3β-interacting regions of these two proteins were shown to possess negligible sequence similarity [126]. Furthermore, even although both bound peptides are primarily helical, their detailed structures and interactions with GSK3β have substantial differences. It has been emphasized that the ability of GSK3β to bind two different proteins with high specificity via the same binding site is mediated by the conformational plasticity of the 285–299 loop. In fact, in the non-bound GSK3β, this loop is highly mobile and poorly defined. However, it is induced to accommodate one of the two completely distinct and well-ordered structures that are specific to the individual ligand bound: while some residues in this versatile binding site in GSK3β are involved in interactions with both axin and FRAT, others are involved uniquely with one or the other [126].

Among other advantages of intrinsic lack of structure and function-related disorder-order transitions are [33, 35, 46, 50, 127, 128]: (*i*) The capability to overcome steric restrictions, enabling essentially larger interaction surfaces in the complex than could be obtained for the rigid partners; (*ii*) The precise control and simple regulation of the binding thermodynamics; and (*iii*) The reduced life-time of intrinsically disordered proteins in the cell, possibly representing a mechanism of rapid turnover of the important regulatory molecules.

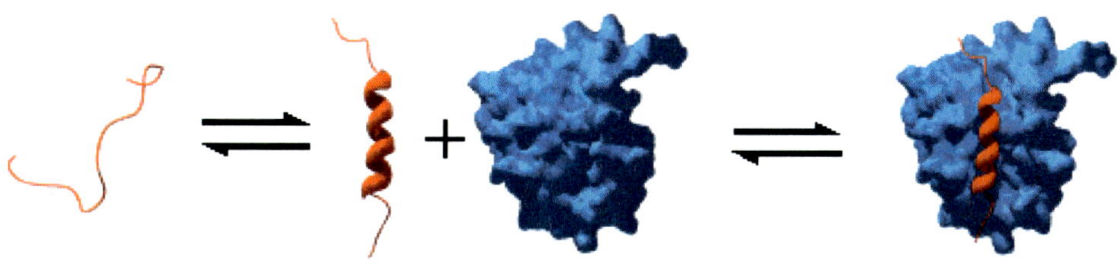

Figure 7. Illustration of disorder-to-order transition upon binding. This example shows the binding of a disordered region of Bad (red ribbon) to Bcl-$X_L$ (blue globular structure).

## Importance of Disorder-to-Order Transitions

Molecular recognition and binding functions can be found within three broad classes: effector, scavenger, and assembler [36]. In each of these cases, a significant disorder-to-order transition occurs upon binding to the target molecule. Some illustrative examples of this coupled folding and binding include the following: cation binding to α-synuclein [129]; self-association of an inhibitory subunit of phosphodiesterase [130]; formation of a functioning ribosome [131] from disordered ribosomal proteins [19]; zinc-induced folding of the DNA-binding domain of the 1,25-dihydroxyvitamin D3 receptor [132]; the SNARE complex formation from the disordered components, Snc1 and Sec9 [133]; interaction of intrinsically disordered caldesmon with its binding partner calmodulin [134]. Many other protein-protein and protein-nucleic acid interactions involve coupled folding and binding of at least one of the partners [33-36, 42, 88, 89, 99-104]. Obviously, the amino acid compositions and the structures of such natively disordered proteins enable a very large accessible surface area (ASA) in the absence of binding partner(s), whereas the binding-induced disorder-to-order transition is accompanied by a dramatic decrease in ASA and also by the concomitant release of a large number of water molecules. In agreement with this hypothesis the binding of lac repressor to its regulatory site is accompanied by the release of ~260 water molecules, as estimated by the osmotic stress method introduced by Parsegian and co-workers [135-137], which is based on the measurement of the change in the number of water molecules present in system compartments that are inaccessible to neutral solutes [138]. Thus, the lack of ability to decrease ASA intramolecularly (i.e., to fold) is compensated in many ID proteins by their outstanding capability to do so upon interaction with their natural partners.

The function-associated conformational changes and disorder-to-order transition may be brought about by alterations in environmental or cellular conditions. This process is illustrated in Figure 7, which shows how the disordered region folds into an α-helical segment upon interaction with the binding partner. The importance of this mechanism originates from the following simple reasoning: a large decrease in conformation entropy, which accompanies disorder-to-order transition, uncouples specificity from binding strength. This phenomenon has the effect of making highly specific interactions easily reversible, which is beneficial for cells, especially in the inducible responses typically involved in signaling and regulation. A recent computational study of such binding illustrated that the disordered partner contains a "conformational preference" for the structure it will take upon binding, and that these so-called "preformed elements" tend to be helices [139]. This research validates previous findings for individual protein-protein interactions, such as $p27^{Kip1}$ [140, 141] and p53 [142], both of which have disordered regions with significant helical character that form α-helices upon binding to their partners.

It has been recently shown that it is possible to predict where at least some of these disorder-to-order transitions will occur [59]. In fact, it has been noticed long ago that PONDR® VL-XT sometimes gives short regions of predicted order bounded by regions of predicted disorder [58]. This was first noticed in the 4E binding protein one (4EBP1), which NMR studies had shown to be completely disordered [143]. However, a short stretch of 4EBP1 undergoes a disorder-to-order transition upon binding to eukaryotic translation initiation factor 4E [144]. The activity of eukaryotic translation initiation factor 4E is inhibited by this binding. The structure of this complex is shown in Figure 8A, whereas the

PONDR® VL-XT prediction for 4EBP1 is shown in Figure 8B. Note that there is a sharp dip in the PONDR® VL-XT score in the area of the binding region. This drop is flanked by long regions of predicted disorder. A similar situation exists in a case of calcineurin, a calcium and calmodulin dependent protein phosphatase, which is important in many signaling pathways. The crystal structure of calcineurin is shown in Figure 8C. The A subunit's catalytic domain is followed by a stretch of 95 residues invisible in the x-ray crystal, a short (18 residue) helix spanning the autoinhibitory domain, and another stretch of 35 residues missing from the X-ray crystal [145]. The PONDR® VL-XT prediction of most of this region is shown in Figure 8D, which illustrates that the region of predicted order overlaps with the autoinhibitory helix.

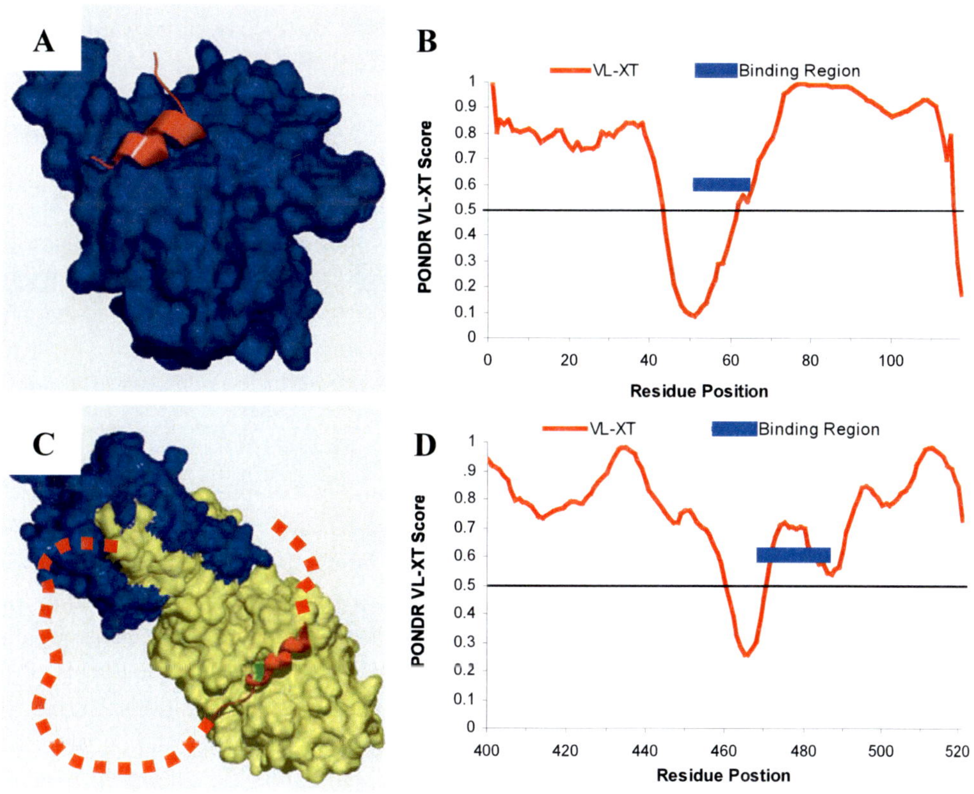


Figure 8. Examples of binding regions and their positions relative to PONDR® predicted order. A. Eukaryotic initiation factor (blue) and the binding region of 4EBP1 (red). **B.** The PONDR® VL-XT prediction for 4EBP1 with the binding region designated (blue bar). **C.** The B (blue) and A (yellow) subunits of calcineurin and the autoinhibitory region of the A subunit (red helix) in the midst of observed disordered sequence (red dashes). **D.** The PONDR® VL-XT prediction for the last 121 amino acid residues of the A subunit with the autoinhibitory region indicated (blue bar). Modified from [59].

Studies of Luisi and colleagues on the *E. coli* ribonuclease RNase E using a combination of several biophysical techniques revealed that the C-terminal half of the protein, which is involved in protein-protein interactions, is intrinsically disordered. Application of PONDR® VL-XT confirmed this conclusion, as this domain was predicted to be entirely disordered except for four sharp downward spikes, which were termed "regions of increased structural propensity" (RISPs) and labeled A, B, C and D [146]. In independent experiments, the functionality of these RISPs was analyzed by synthesizing peptide fragments consisting of each of the RISP regions and studying their biological function. Importantly, these peptide fragments were able to bind to their partners: RISPs C and D were shown to bind to enolase and PNPase, respectively, whereas RISP B was shown to contribute to RNA binding and RISP A was involved into the formation of a coiled-coil site [146]. Furthermore, the crystal structure of the enolase-RISP-C/enolase complex has been determined, where RISP C was shown to form an α-helix (Luisi, personal communication). These findings clearly show that short regions of predicted order bounded by extended regions of predicted disorder in PONDR® VL-XT plots can be used to identify binding sites that involved disorder-to-order transitions upon complex formation.

Additional work has further validated the use of these distinctive downward spikes in PONDR® VL-XT curves to locate functional binding regions [59]. Based on the above-described and similar observations, we propose a protein structural element or feature that mediates the binding events of initially disordered regions. This element consists of a short region that undergoes coupled binding and folding within a longer region of disorder. We call these features "molecular recognition elements," (MoREs). Alternatively, these regions can be named Molecular Recognition Features (MoRFs), to emphasize their unique polymorphic nature. An algorithm has been recently elaborated [59] that identifies regions having a propensity for of α-helix-forming molecular recognition features (α-MoRFs) based on a discriminant function that indicates such regions while giving a low false-positive error rate on a collection of structured proteins. Application of this predictor to databases of genomics and functionally annotated proteins indicates that α-MoREs are likely to be play important roles protein-protein interactions involved in signaling events [59].

Systematic studies of PDB entries revealed that protein complexes deposited in this database often comprise a short peptide bound to a larger globular protein. Analysis of literature data showed that some of these short peptides, being specifically folded (in a form of α-helix, β-hairpin, β-strand, polyproline II helix, or irregular structure, etc.) within their complexes with globular partners, are intrinsically disordered prior the corresponding complex formation. Thus, all these regions can be considered as illustrative members of the subset of protein-protein interactions involving disorder-to-order transitions during the complex formation, known as MORFs. In our recent study, 2,512 polypeptide chains were extracted from PDB, which satisfied the prerequisite of being short (lengths ≤ 70 residues) and bound to a globular partner (with chains ≥ 100 residues). As the next step, all sequences having any occurrences of 'X' and 'Z' and being shorter than 10 residues were removed. This preprocessing of data resulted in a dataset comprising of ~1,200 chains (~55,000 residues and an average chain length of 33.5 residues). This was followed up by the removal of redundancy amongst these ~1200 chains. The final dataset included 372 non-redundant protein chains (9,093 residues). The secondary structure assignment showed that, 27% of this dataset consisted of α-helical residues, 12% were β-sheet residues and approximately 48% of the residues had an irregular conformation. The remaining 13% of the residues were found to be disordered as they were characterized by missing coordinate information in their respective PDB files [147,189]. We believe that these collections of experimentally proven α-, β-, and ι-MoRFs represent useful datasets that can be used to find sequence attributes for discriminating these different MoRFs from each other and from ordered proteins.

## What is the Relationship between ID Proteins and Human Diseases

Because of the fact that ID protein play crucial roles in numerous biological processes, it was not too surprising to find that some of them are involved in human diseases. For example, a number of human diseases originate from the deposition of stable, ordered, filamentous protein aggregates, commonly referred to as amyloid fibrils. In each of these pathological states, a specific protein or protein fragment changes from its natural soluble form into insoluble fibrils, which accumulate in a variety of organs and tissues [148-154]. Approximately 20 different proteins are known so far to be involved in these diseases. These proteins are unrelated in terms of sequence or starting structure. Several ID proteins are found in this list of 20, being associated with the development of several neurodegenerative diseases [154, 155]. An incomplete list of disordered associated with ID proteins includes Alzheimer's disease (deposition of amyloid-β, tau-protein, α-synuclein fragment NAC [156-159]; Niemann-Pick disease type C, subacute sclerosing panencephalitis, argyrophilic grain disease, myotonic dystrophy, and motor neuron disease with neurofibrillary tangles (accumulation of tau-protein in form of neurofibrillary tangles [158]); Down's syndrome (nonfilamentous amyloid-β deposits [160]); Parkinson's disease, dementia with Lewy body, diffuse Lewy body disease, Lewy body variant of Alzheimer's disease, multiple system atrophy and Hallervorden-Spatz disease (deposition of α-synuclein in a form of Lewy body, or Lewy neuritis [161]); prion diseases (deposition of $PrP^{SC}$ [162]); and a family of polyQ diseases, a group of neurodegenerative disorders caused by expansion of GAC trinucleotide repeats coding for PolyQ in the gene products [163]. Furthermore, most mutations in rigid globular proteins associated with accelerated fibrillation and protein deposition diseases have been shown to destabilize the native structure, increasing the steady-state concentration of partially folded (disordered) conformers [148-154].

The disorders given above have been called conformational diseases, as they are characterized by the conformational changes, misfolding and aggregation of an underlying protein. However, there is another side to this coin: protein functionality. In fact, many of the proteins associated with the conformational disorders are also involved in recognition, regulation and cell signaling. For example, functions ascribed to α-synuclein, a protein involved in several neurodegenerative disorders, include binding fatty acids and metal ions; regulation of certain enzymes, transporters, and neurotransmitter vesicles; and regulation of neuronal survival (reviewed in [161]). Overall, there are about 50 (!) proteins and ligands that interact and/or co-localize with this protein. Furthermore, α-synuclein has amazing structural plasticity and adopts a series of different monomeric, oligomeric and insoluble conformations (reviewed in [164]). The choice between these conformations is determined by the peculiarities of the protein environment, assuming that α-synuclein has an exceptional ability to fold in a template-dependent manner. Based on these observations, we hypothesize that the development of the conformational diseases may originate from the misidentification, misregulation and missignaling, accompanied by misfolding. In other words, mutations and/or changes in the environment may result in protein confusion, for which its ID becomes lost, thus reducing its capability to recognize proper binding partners and leading to the formation of nonfunctional and deadly aggregates.

Recent analysis of so-called polyglytamine diseases gives support to this hypothesis [165]. Polyglytamine diseases are a specific group of hereditary neurodegeneration caused by expansion of CAG triplet repeats in an exon of disease genes which leads to the production of a disease protein containing an expanded polyglutamine, polyQ, stretch. Nine neurodegenerative disorders, including Kennedy's disease, Huntington's diseases, spinocerebellar atrophy-1, -2, -3, -6, 7, 17, and dentatorubral pallidoluysian atrophy are known to belong to this class of diseases [166-169]. In most polyQ diseases, expansion to over 40 repeats leads to the onset [169]. It has been emphasized that such molecular processes as unfolded protein response, protein transport, synaptic transmission and transcription are implicated in the pathology of polyQ diseases [165]. Importantly, more than 20 transcription-related factors have been reported to interact with pathological polyQ proteins. Furthermore, these interactions were shown to repress the transcription, leading finally to the neuronal dysfunction and death (reviewed in [165]). These results suggest that polyQ diseases represent kind of transcriptional disorder [165], supporting our misidentification hypothesis for at least some of the conformational disorders.

Disorder is very common in cancer-associated proteins too. In a 2002 study, it was found that 79% of cancer-associated and 66% of cell-signaling proteins contain predicted regions of disorder of 30 residues or longer [42]. In contrast, only 13% of a set of proteins with well-defined ordered structures contained such long regions of predicted disorder. For this study, cancer-associated proteins were defined as those human proteins in Swiss-Prot containing the keyword "oncogene" (this included anti- and proto-oncogenes) or containing the word "tumor" in the description field. In experimental studies, the presence of disorder has been directly observed in several cancer-associated proteins, including p53 [142], $p57^{kip2}$ [170], Bcl-$X_L$ and Bcl-2 [171], c-Fos [172], and most recently, a thyroid cancer associated protein, TC-1 [173].

# Conclusion

Intrinsically disordered (ID) proteins are widespread and represent a distinct protein tribe, with disorder being an important structural element that exists at various levels of protein structure. Such ID proteins are commonly involved in recognition, regulation and cell signaling functions and have biophysical characteristics that are well disposed for this role. They are much more common in eukaryota in comparison to prokaryota and archaea, reflecting the greater need for disorder-associated signaling and regulation in nucleated cells. Changes in the environment and/or mutation(s) would be expected to affect the normal function of the ID proteins, leading to misidentification and missignaling. This, in turn, can result in misfolding and aggregation, which are known to be associated with the pathogenesis of numerous disease states.

## Acknowledgements

We thank Zoran Obradovic and Ya-Yue Van for their continuing support and collaborations in the field of the ID proteins studies. We also thank T. Sosnick for assistance in generating the depicted U, pre-MG and MG states. This work was supported in part by the NIH grant

R01 LM007688-01A1 (to A. K. D.) and the Programs of the Russian Academy of Sciences for the "Molecular and cellular biology" and "Fundamental science for medicine" (to V. N. U.).

## References

[1] Fischer, E. (1894) Einfluss der configuration auf die wirkung der enzyme. *Ber. Dt. Chem. Ges.* **27**, 2985-2993.

[2] Lemieux, U. R. & Spohr, U. (1994) How emil fischer was led to the lock and key concept for enzyme specificity. *Adv. Carbohydrate Chem. Biochem.* **50**, 1-20.

[3] Blake, C. C., Koenig, D. F., Mair, G. A., North, A. C., Phillips, D. C. & Sarma, V. R. (1965) Structure of hen egg-white lysozyme. A three-dimensional fourier synthesis at 2 angstrom resolution. *Nature* **206**, 757-761.

[4] Kendrew, J. C., Bodo, G., Dintzis, H. M., Parrish, R. G., Wyckoff, H. & Phillips, D. C. (1958) A three-dimensional model of the myoglobin molecule obtained by x-ray analysis. *Nature* **181**, 662-666.

[5] Kendrew, J. C., Dickerson, R. E., Stranberg, B. E., R.J., H., Davies, D. R., Phillips, D. C. & Shore, V. C. (1960) Structure of myoglobin: A three-dimensional fourier synthesis at 2 å resolution. *Nature* **185**, 422-427.

[6] Bernstein, F. C., Koetzle, T. F., Williams, G. J., Meyer, E. F., Jr., Brice, M. D., Rodgers, J. R., Kennard, O., Shimanouchi, T. & Tasumi, M. (1977) The protein data bank: A computer-based archival file for macromolecular structures. *J Mol Biol* **112**, 535-542.

[7] Petsko, G. A. & Ringe, D. (2004) *Primers in biology. Protein structure and function.*, New Science Press Ltd., Sinauer Associates, Inc. Publishers, Blackwell Publishing, London.

[8] Bloomer, A. C., Champness, J. N., Bricogne, G., Staden, R. & Klug, A. (1978) Protein disk of tobacco mosaic virus at 2.8 a resolution showing the interactions within and between subunits. *Nature* **276**, 362-368.

[9] Bode, W., Schwager, P. & Huber, R. (1978) The transition of bovine trypsinogen to a trypsin-like state upon strong ligand binding. The refined crystal structures of the bovine trypsinogen-pancreatic trypsin inhibitor complex and of its ternary complex with ile-val at 1.9 a resolution. *J Mol Biol* **118**, 99-112.

[10] Sedzik, J. & Kirschner, D. A. (1992) Is myelin basic protein crystallizable? *Neurochem Res* **17**, 157-166.

[11] Harauz, G., Ishiyama, N., Hill, C. M., Bates, I. R., Libich, D. S. & Fares, C. (2004) Myelin basic protein-diverse conformational states of an intrinsically unstructured protein and its roles in myelin assembly and multiple sclerosis. *Micron* **35**, 503-542.

[12] Lian, L. Y. (1998) Nmr structural studies of glutathione s-transferase. *Cell Mol Life Sci* **54**, 359-362.

[13] Sidote, D. J. & Hoffman, D. W. (2003) Nmr structure of an archaeal homologue of ribonuclease p protein rpp29. *Biochemistry* **42**, 13541-13550.

[14] Karush, F. (1950) Heterogeneity of the binding sites of bovine serum albumin. *J. Am. Chem. Soc.* **72**, 2705-2713.

[15] Watts, J. D., Cary, P. D., Sautiere, P. & Crane-Robinson, C. (1990) Thymosins: Both nuclear and cytoplasmic proteins. *Eur J Biochem* **192**, 643-651.

[16] Gast, K., Damaschun, H., Eckert, K., Schulze-Forster, K., Maurer, H. R., Muller-Frohne, M., Zirwer, D., Czarnecki, J. & Damaschun, G. (1995) Prothymosin alpha: A biologically active protein with random coil conformation. *Biochemistry* **34**, 13211-13218.

[17] Uversky, V. N., Gillespie, J. R., Millett, I. S., Khodyakova, A. V., Vasiliev, A. M., Chernovskaya, T. V., Vasilenko, R. N., Kozlovskaya, G. D., Dolgikh, D. A., Fink, A. L., Doniach, S. & Abramov, V. M. (1999) Natively unfolded human prothymosin alpha adopts partially folded collapsed conformation at acidic ph. *Biochemistry* **38**, 15009-15016.

[18] Boublik, M., Bradbury, E. M., Crane-Robinson, C. & Johns, E. W. (1970) An investigation of the conformational changes of histone f2b by high resolution nuclear magnetic resonance. *Eur J Biochem* **17**, 151-159.

[19] Venyaminov, S. Y., Gudkov, A. T., Gogia, Z. V. & Tumanova, L. G. (1981) *Absorption and circular dichroism spectra of individual proteins from escherichia coli ribosomes*, Pushchino, Russia.

[20] Breslow, E., Beychok, S., Hardman, K. D. & Gurd, F. R. (1965) Relative conformations of sperm whale metmyoglobin and apomyoglobin in solution. *J Biol Chem* **240**, 304-309.

[21] Stellwagen, E., Rysavy, R. & Babul, G. (1972) The conformation of horse heart apocytochrome c. *J Biol Chem* **247**, 8074-8077.

[22] Fisher, W. R., Taniuchi, H. & Anfinsen, C. B. (1973) On the role of heme in the formation of the structure of cytochrome c. *J Biol Chem* **248**, 3188-3195.

[23] Isbell, D. T., Du, S., Schroering, A. G., Colombo, G. & Shelling, J. G. (1993) Metal ion binding to dog osteocalcin studied by 1h nmr spectroscopy. *Biochemistry* **32**, 11352-11362.

[24] Bychkova, V. E., Pain, R. H. & Ptitsyn, O. B. (1988) The 'molten globule' state is involved in the translocation of proteins across membranes? *FEBS Lett* **238**, 231-234.

[25] Bychkova, V. E. & Ptitsyn, O. B. (1993) The molten globule in vitro and in vivo. *Chemtracts Biochem. Molec. Biol.* **4**, 133-163.

[26] Ptitsyn, O. B., Bychkova, V. E. & Uversky, V. N. (1995) Kinetic and equilibrium folding intermediates. *Philos Trans R Soc Lond B Biol Sci* **348**, 35-41.

[27] Martin, J., Langer, T., Boteva, R., Schramel, A., Horwich, A. L. & Hartl, F. U. (1991) Chaperonin-mediated protein folding at the surface of groel through a 'molten globule'-like intermediate. *Nature* **352**, 36-42.

[28] van der Goot, F. G., Gonzalez-Manas, J. M., Lakey, J. H. & Pattus, F. (1991) A 'molten-globule' membrane-insertion intermediate of the pore-forming domain of colicin a. *Nature* **354**, 408-410.

[29] van der Goot, F. G., Lakey, J. H. & Pattus, F. (1992) The molten globule intermediate for protein insertion or translocation through membranes. *Trends Cell Biol* **2,** 343-348.

[30] Uversky, V. N. & Narizhneva, N. V. (1998) Effect of natural ligands on the structural properties and conformational stability of proteins. *Biochemistry (Mosc)* **63**, 420-433.

[31] Uversky, V. N. (2003) A rigidifying union: The role of ligands in protein structure and stability. in *Recent research developments in biophysics & biochemistry* (Pandalai, S. G., Ed.) pp 711-745, Transworld Research Network, Kerala, India.

[32] Bychkova, V. E. & Ptitsyn, O. B. (1995) Folding intermediates are involved in genetic diseases? *FEBS Lett* **359**, 6-8.

[33] Wright, P. E. & Dyson, H. J. (1999) Intrinsically unstructured proteins: Re-assessing the protein structure-function paradigm. *J Mol Biol* **293**, 321-331.

[34] Uversky, V. N., Gillespie, J. R. & Fink, A. L. (2000) Why are "natively unfolded" proteins unstructured under physiologic conditions? *Proteins* **41**, 415-427.

[35] Dunker, A. K., Lawson, J. D., Brown, C. J., Williams, R. M., Romero, P., Oh, J. S., Oldfield, C. J., Campen, A. M., Ratliff, C. M., Hipps, K. W., Ausio, J., Nissen, M. S., Reeves, R., Kang, C., Kissinger, C. R., Bailey, R. W., Griswold, M. D., Chiu, W., Garner, E. C. & Obradovic, Z. (2001) Intrinsically disordered protein. *J Mol Graph Model* **19**, 26-59.

[36] Tompa, P. (2002) Intrinsically unstructured proteins. *Trends Biochem Sci 27*, 527-533.

[37] Schweers, O., Schonbrunn-Hanebeck, E., Marx, A. & Mandelkow, E. (1994) Structural studies of tau protein and alzheimer paired helical filaments show no evidence for beta-structure. *J Biol Chem* **269**, 24290-24297.

[38] Weinreb, P. H., Zhen, W., Poon, A. W., Conway, K. A. & Lansbury, P. T., Jr. (1996) Nacp, a protein implicated in alzheimer's disease and learning, is natively unfolded. *Biochemistry* **35**, 13709-13715.

[39] Daughdrill, G. W., Pielak, G. J., Uversky, V. N., Cortese, M. S. & Dunker, A. K. (2005) Natively disordered proteins. in *Handbook of protein folding* (Buchner, J. & Kiefhaber, T., Eds.) pp 271-353, Wiley-VCH, Verlag GmbH & Co., Weinheim, Germany.

[40] Oldfield, C. J., Cheng, Y., Cortese, M. S., Brown, C. J., Uversky, V. N. & Dunker, A. K. (2005) Comparing and combining predictors of mostly disordered proteins. *Biochemistry* **44**, 1989-2000.

[41] Dunker, A. K., Obradovic, Z., Romero, P., Garner, E. C. & Brown, C. J. (2000) Intrinsic protein disorder in complete genomes. *Genome Inform Ser Workshop Genome Inform* **11**, 161-171.

[42] Iakoucheva, L. M., Brown, C. J., Lawson, J. D., Obradovic, Z. & Dunker, A. K. (2002) Intrinsic disorder in cell-signaling and cancer-associated proteins. *J Mol Biol* **323**, 573-584.

[43] Dunker, A. K. & Obradovic, Z. (2001) The protein trinity--linking function and disorder. *Nat Biotechnol* **19**, 805-806.

[44] Hemmings, H. C., Jr., Nairn, A. C., Aswad, D. W. & Greengard, P. (1984) Darpp-32, a dopamine- and adenosine 3':5'-monophosphate-regulated phosphoprotein enriched in dopamine-innervated brain regions. Ii. Purification and characterization of the phosphoprotein from bovine caudate nucleus. *J Neurosci* **4,** 99-110.

[45] Williams, R. J. (1978) The conformational mobility of proteins and its functional significance. *Biochem Soc Trans* **6,** 1123-1126.

[46] Dunker, A. K., Garner, E., Guilliot, S., Romero, P., Albrecht, K., Hart, J., Obradovic, Z., Kissinger, C. & Villafranca, J. E. (1998) Protein disorder and the evolution of molecular recognition: Theory, predictions and observations. *Pac Symp Biocomput*, 473-484.

[47] Garner, E., Cannon, P., Romero, P., Obradovic, Z. & Dunker, A. K. (1998) Predicting disordered regions from amino acid sequence: Common themes despite differing structural characterization. *Genome Inform Ser Workshop Genome Inform* **9**, 201-213.

[48] Vihinen, M., Torkkila, E. & Riikonen, P. (1994) Accuracy of protein flexibility predictions. *Proteins* **19**, 141-149.

[49] Williams, R. M., Obradovi, Z., Mathura, V., Braun, W., Garner, E. C., Young, J., Takayama, S., Brown, C. J. & Dunker, A. K. (2001) The protein non-folding problem: Amino acid determinants of intrinsic order and disorder. *Pac Symp Biocomput*, 89-100.

[50] Romero, P., Obradovic, Z., Li, X., Garner, E. C., Brown, C. J. & Dunker, A. K. (2001) Sequence complexity of disordered protein. *Proteins* **42**, 38-48.

[51] Li, X., Obradovic, Z., Brown, C. J., Garner, E. C. & Dunker, A. K. (2000) Comparing predictors of disordered protein. *Genome Inform Ser Workshop Genome Inform* **11**, 172-184.

[52] Romero, P., Obradovic, Z., Kissinger, C. R., Villafranca, J. E. & Dunker, A. K. (1997) Identifying disordered regions in proteins from amino acid sequences. *IEEE Int. Conf. Neural Netw.* **1**, 90-95.

[53] Romero, P., Obradovic, Z. & Dunker, A. K. (2000) Intelligent data analysis for protein disorder prediction. *Artif. Intel. Rev.* **14**, 447-484.

[54] Li, X., Romero, P., Rani, M., Dunker, A. K. & Obradovic, Z. (1999) Predicting protein disorder for n-, c-, and internal regions. *Genome Inform Ser Workshop Genome Inform* **10**, 30-40.

[55] Romero, Obradovic & Dunker, K. (1997) Sequence data analysis for long disordered regions prediction in the calcineurin family. *Genome Inform Ser Workshop Genome Inform* **8,** 110-124.

[56] Vucetic, S., Radivojac, P., Obradovic, Z., Brown, C. J. & Dunker, A. K. (2001) Methods for improving protein disorder prediction. in *Ieee int. Conf. Neural netw*, Washington, DC.

[57] Peng, K., Vucetic, S., Radivojac, P., Brown, C. J., Dunker, A. K. & Obradovic, Z. (2005) Optimizing long intrinsic disorder predictors with protein evolutionary information. *J Bioinform Comput Biol* **3,** 35-60.

[58] Garner, E., Romero, P., Dunker, A. K., Brown, C. & Obradovic, Z. (1999) Predicting binding regions within disordered proteins. *Genome Inform Ser Workshop Genome Inform* **10**, 41-50.

[59] Oldfield, C. J., Cheng, Y., Cortese, M. S., Romero, P., Uversky, V. N. & Dunker, A. K. (2005) Coupled folding and binding with alpha-helix-forming molecular recognition elements. *Biochemistry*, In press.

[60] Melamud, E. & Moult, J. (2003) Evaluation of disorder predictions in casp5. *Proteins* **53** Suppl 6, 561-565.

[61] Liu, J., Tan, H. & Rost, B. (2002) Loopy proteins appear conserved in evolution. *J Mol Biol* **322**, 53-64.

[62] Vucetic, S., Brown, C. J., Dunker, A. K. & Obradovic, Z. (2003) Flavors of protein disorder. *Proteins* **52**, 573-584.

[63] Uversky, V. N. & Ptitsyn, O. B. (1994) "partly folded" state, a new equilibrium state of protein molecules: Four-state guanidinium chloride-induced unfolding of beta-lactamase at low temperature. *Biochemistry* **33**, 2782-2791.

[64] Fink, A. L. (1995) Compact intermediate states in protein folding. *Annu Rev Biophys Biomol Struct* **24**, 495-522.

[65] Uversky, V. N. & Ptitsyn, O. B. (1996) Further evidence on the equilibrium "pre-molten globule state": Four-state guanidinium chloride-induced unfolding of carbonic anhydrase b at low temperature. *J Mol Biol* **255**, 215-228.
[66] Uversky, V. N. (2003) Protein folding revisited. A polypeptide chain at the folding-misfolding-nonfolding cross-roads: Which way to go? *Cell Mol Life Sci* **60**, 1852-1871.
[67] Ptitsyn, O. B. (1995) Molten globule and protein folding. *Adv Protein Chem* **47**,83-229.
[68] Baum, J., Dobson, C. M., Evans, P. A. & Hanley, C. (1989) Characterization of a partly folded protein by nmr methods: Studies on the molten globule state of guinea pig alpha-lactalbumin. *Biochemistry* **28**, 7-13.
[69] Bushnell, G. W., Louie, G. V. & Brayer, G. D. (1990) High-resolution three-dimensional structure of horse heart cytochrome c. *J Mol Biol* **214**, 585-595.
[70] Chyan, C. L., Wormald, C., Dobson, C. M., Evans, P. A. & Baum, J. (1993) Structure and stability of the molten globule state of guinea-pig alpha-lactalbumin: A hydrogen exchange study. *Biochemistry* **32**, 5681-5691.
[71] Jeng, M. F., Englander, S. W., Elove, G. A., Wand, A. J. & Roder, H. (1990) Structural description of acid-denatured cytochrome c by hydrogen exchange and 2d nmr. *Biochemistry* **29**, 10433-10437.
[72] Wu, L. C., Laub, P. B., Elove, G. A., Carey, J. & Roder, H. (1993) A noncovalent peptide complex as a model for an early folding intermediate of cytochrome c. *Biochemistry* **32**, 10271-10276.
[73] Eliezer, D., Yao, J., Dyson, H. J. & Wright, P. E. (1998) Structural and dynamic characterization of partially folded states of apomyoglobin and implications for protein folding. *Nat Struct Biol* **5**, 148-155.
[74] Bose, H. S., Whittal, R. M., Baldwin, M. A. & Miller, W. L. (1999) The active form of the steroidogenic acute regulatory protein, star, appears to be a molten globule. *Proc Natl Acad Sci U S A* **96**, 7250-7255.
[75] Bracken, C. (2001) Nmr spin relaxation methods for characterization of disorder and folding in proteins. *J Mol Graph Model* **19**, 3-12.
[76] Merrill, A. R., Cohen, F. S. & Cramer, W. A. (1990) On the nature of the structural change of the colicin e1 channel peptide necessary for its translocation-competent state. *Biochemistry* **29**, 5829-5836.
[77] Fontana, A., Fassina, G., Vita, C., Dalzoppo, D., Zamai, M. & Zambonin, M. (1986) Correlation between sites of limited proteolysis and segmental mobility in thermolysin. *Biochemistry* **25**, 1847-1851.
[78] Fontana, A., Polverino de Laureto, P. & De Phillips, V. (1993) Molecular aspects of proteolysis of globular proteins. in *Protein stability and stabilization* (van den Tweel, W., Harder, A. & Buitelear, M., Eds.) pp 101-110, Elsevier Science, Amsterdam, The Netherlands.
[79] Fontana, A., Zambonin, M., Polverino de Laureto, P., De Filippis, V., Clementi, A. & Scaramella, E. (1997) Probing the conformational state of apomyoglobin by limited proteolysis. *J Mol Biol* **266**, 223-230.
[80] Fontana, A., Polverino de Laureto, P., De Filippis, V., Scaramella, E. & Zambonin, M. (1997) Probing the partly folded states of proteins by limited proteolysis. *Fold Des* **2**, R17-26.

[81] Semisotnov, G. V., Rodionova, N. A., Razgulyaev, O. I., Uversky, V. N., Gripas, A. F. & Gilmanshin, R. I. (1991) Study of the "molten globule" intermediate state in protein folding by a hydrophobic fluorescent probe. *Biopolymers* **31**, 119-128.

[82] Uversky, V. N. (1993) Use of fast protein size-exclusion liquid chromatography to study the unfolding of proteins which denature through the molten globule. *Biochemistry* **32**, 13288-13298.

[83] Eliezer, D., Chiba, K., Tsuruta, H., Doniach, S., Hodgson, K. O. & Kihara, H. (1993) Evidence of an associative intermediate on the myoglobin refolding pathway. *Biophys J* **65**, 912-917.

[84] Kataoka, M., Kuwajima, K., Tokunaga, F. & Goto, Y. (1997) Structural characterization of the molten globule of alpha-lactalbumin by solution x-ray scattering. *Protein Sci* **6,** 422-430.

[85] Semisotnov, G. V., Kihara, H., Kotova, N. V., Kimura, K., Amemiya, Y., Wakabayashi, K., Serdyuk, I. N., Timchenko, A. A., Chiba, K., Nikaido, K., Ikura, T. & Kuwajima, K. (1996) Protein globularization during folding. A study by synchrotron small-angle x-ray scattering. *J Mol Biol* **262**, 559-574.

[86] Uversky, V. N., Karnoup, A. S., Segel, D. J., Seshadri, S., Doniach, S. & Fink, A. L. (1998) Anion-induced folding of staphylococcal nuclease: Characterization of multiple equilibrium partially folded intermediates. *J Mol Biol* **278**, 879-894.

[87] Tcherkasskaya, O. & Uversky, V. N. (2001) Denatured collapsed states in protein folding: Example of apomyoglobin. *Proteins* **44**, 244-254.

[88] Uversky, V. N. (2002) Natively unfolded proteins: A point where biology waits for physics. *Protein Sci* **11**, 739-756.

[89] Uversky, V. N. (2002) What does it mean to be natively unfolded? *Eur J Biochem* **269**, 2-12.

[90] Hammarstrom, P. & Carlsson, U. (2000) Is the unfolded state the rosetta stone of the protein folding problem? *Biochem Biophys Res Commun* **276**, 393-398.

[91] Shortle, D. (2002) The expanded denatured state: An ensemble of conformations trapped in a locally encoded topological space. *Adv Protein Chem* **62**, 1-23.

[92] Smith, L. J., Fiebig, K. M., Schwalbe, H. & Dobson, C. M. (1996) The concept of a random coil. Residual structure in peptides and denatured proteins. *Fold Des* **1,** R95-106.

[93] Shimizu, S. & Chan, H. S. (2002) Origins of protein denatured state compactness and hydrophobic clustering in aqueous urea: Inferences from nonpolar potentials of mean force. *Proteins* **49**, 560-566.

[94] Uverskii, V. N. (1998) How many molten globules states exist? *Biofizika* **43**, 416-421.

[95] Tcherkasskaya, O. & Uversky, V. N. (2003) Polymeric aspects of protein folding: A brief overview. *Protein Pept Lett* **10**, 239-245.

[96] Tcherkasskaya, O., Davidson, E. A. & Uversky, V. N. (2003) Biophysical constraints for protein structure prediction. *J Proteome Res* **2**, 37-42.

[97] Cortese, M. S., Baird, J. P., Uversky, V. N. & Dunker, A. K. (2005) Uncovering the unfoldome: Enriching cell extracts for unstructured proteins by acid treatment. *J Proteome Res* **4,** 1610-1618.

[98] Iakoucheva, L. M., Kimzey, A. L., Masselon, C. D., Smith, R. D., Dunker, A. K. & Ackerman, E. J. (2001) Aberrant mobility phenomena of the DNA repair protein xpa. *Protein Sci* **10**, 1353-1362.

[99] Dunker, A. K., Brown, C. J., Lawson, J. D., Iakoucheva, L. M. & Obradovic, Z. (2002) Intrinsic disorder and protein function. *Biochemistry* **41**, 6573-6582.

[100] Dyson, H. J. & Wright, P. E. (2002) Coupling of folding and binding for unstructured proteins. *Curr Opin Struct Biol* **12**, 54-60.

[101] Dyson, H. J. & Wright, P. E. (2005) Intrinsically unstructured proteins and their functions. *Nat Rev Mol Cell Biol* **6**, 197-208.

[102] Fink, A. L. (2005) Natively unfolded proteins. *Curr Opin Struct Biol* **15**, 35-41.

[103] Uversky, V. N., Oldfield, C. J. & Dunker, A. K. (2005) Showing your id: Intrinsic disorder as an id for recognition, regulation and cell signaling. *J Mol Recognit* **18**, 343-384.

[104] Dunker, A. K., Cortese, M. S., Romero, P., Iakoucheva, L. M. & Uversky, V. N. (2005) Flexible nets: The roles of intrinsic disorder in protein interaction networks. *FEBS Journal* **272**, 5129-5148.

[105] Dunker, A. K., Brown, C. J. & Obradovic, Z. (2002) Identification and functions of usefully disordered proteins. *Adv Protein Chem* **62**, 25-49.

[106] Antz, C., Geyer, M., Fakler, B., Schott, M. K., Guy, H. R., Frank, R., Ruppersberg, J. P. & Kalbitzer, H. R. (1997) Nmr structure of inactivation gates from mammalian voltage-dependent potassium channels. *Nature* **385**, 272-275.

[107] Armstrong, C. M. & Bezanilla, F. (1977) Inactivation of the sodium channel. Ii. Gating current experiments. *J Gen Physiol* **70**, 567-590.

[108] Herson, P. S., Virk, M., Rustay, N. R., Bond, C. T., Crabbe, J. C., Adelman, J. P. & Maylie, J. (2003) A mouse model of episodic ataxia type-1. *Nat Neurosci* **6**, 378-383.

[109] Lerche, H., Jurkat-Rott, K. & Lehmann-Horn, F. (2001) Ion channels and epilepsy. *Am J Med Genet* **106**, 146-159.

[110] Liebovitch, L. S., Selector, L. Y. & Kline, R. P. (1992) Statistical properties predicted by the ball and chain model of channel inactivation. *Biophys J* **63**, 1579-1585.

[111] Hoshi, T., Zagotta, W. N. & Aldrich, R. W. (1990) Biophysical and molecular mechanisms of shaker potassium channel inactivation. *Science* **250**, 533-538.

[112] Erdös, P. & Rényi, A. (1960) On the evolution of random graphs. *Publ. Math. Inst. Hung. Acad. Sci.* **5**, 17-61.

[113] Barabasi, A. L. & Bonabeau, E. (2003) Scale-free networks. *Sci Am* **288**, 60-69.

[114] Watts, D. J. & Strogatz, S. H. (1998) Collective dynamics of 'small-world' networks. *Nature* **393**, 440-442.

[115] Goh, K. I., Oh, E., Jeong, H., Kahng, B. & Kim, D. (2002) Classification of scale-free networks. *Proc Natl Acad Sci U S A* **99**, 12583-12588.

[116] Barabasi, A. L. & Albert, R. (1999) Emergence of scaling in random networks. *Science* **286**, 509-512.

[117] Koshland, D. E., Jr. (1958) Application of a theory of enzyme specificity to protein synthesis. *Proc Natl Acad Sci USA* **44**, 98-104.

[118] Sawaya, M. R. & Kraut, J. (1997) Loop and subdomain movements in the mechanism of escherichia coli dihydrofolate reductase: Crystallographic evidence. *Biochemistry* **36**, 586-603.

[119] Schulz, G. E. (1979) Nucleotide binding proteins. in *Molecular mechanism of biological recognition* (Balaban, M., Ed.) pp 79-94, Elsevier/North-Holland Biomedical Press, New York.

[120] Kriwacki, R. W., Hengst, L., Tennant, L., Reed, S. I. & Wright, P. E. (1996) Structural studies of p21waf1/cip1/sdi1 in the free and cdk2-bound state: Conformational disorder mediates binding diversity. *Proc Natl Acad Sci U S A* **93**, 11504-11509.

[121] Jeong, H., Mason, S. P., Barabasi, A. L. & Oltvai, Z. N. (2001) Lethality and centrality in protein networks. *Nature* **411**, 41-42.

[122] Iakoucheva, L. M., Kimzey, A. L., Masselon, C. D., Bruce, J. E., Garner, E. C., Brown, C. J., Dunker, A. K., Smith, R. D. & Ackerman, E. J. (2001) Identification of intrinsic order and disorder in the DNA repair protein xpa. *Protein Sci* **10**, 560-571.

[123] Choo, Y. & Schwabe, J. W. (1998) All wrapped up. *Nat Struct Biol* **5**, 253-255.

[124] Meador, W. E., Means, A. R. & Quiocho, F. A. (1992) Target enzyme recognition by calmodulin: 2.4 a structure of a calmodulin-peptide complex. *Science* **257**, 1251-1255.

[125] Pontius, B. W. (1993) Close encounters: Why unstructured, polymeric domains can increase rates of specific macromolecular association. *Trends Biochem Sci* **18**, 181-186.

[126] Dajani, R., Fraser, E., Roe, S. M., Yeo, M., Good, V. M., Thompson, V., Dale, T. C. & Pearl, L. H. (2003) Structural basis for recruitment of glycogen synthase kinase 3beta to the axin-apc scaffold complex. *Embo J* **22**, 494-501.

[127] Dunker, A. K., Obradovic, Z., Romero, P., Kissinger, C. & Villafranca, E. (1997) On the importance of being disordered. *PDB Newsletter* **81**, 3-5.

[128] Brown, C. J., Takayama, S., Campen, A. M., Vise, P., Marshall, T. W., Oldfield, C. J., Williams, C. J. & Dunker, A. K. (2002) Evolutionary rate heterogeneity in proteins with long disordered regions. *J Mol Evol* **55**, 104-110.

[129] Uversky, V. N., Li, J. & Fink, A. L. (2001) Metal-triggered structural transformations, aggregation, and fibrillation of human alpha-synuclein. A possible molecular nk between parkinson's disease and heavy metal exposure. *J Biol Chem* **276**, 44284-44296.

[130] Uversky, V. N., Permyakov, S. E., Zagranichny, V. E., Rodionov, I. L., Fink, A. L., Cherskaya, A. M., Wasserman, L. A. & Permyakov, E. A. (2002) Effect of zinc and temperature on the conformation of the gamma subunit of retinal phosphodiesterase: A natively unfolded protein. *J Proteome Res* **1**, 149-159.

[131] Yusupov, M. M., Yusupova, G. Z., Baucom, A., Lieberman, K., Earnest, T. N., Cate, J. H. & Noller, H. F. (2001) Crystal structure of the ribosome at 5.5 a resolution. *Science* **292**, 883-896.

[132] Craig, T. A., Veenstra, T. D., Naylor, S., Tomlinson, A. J., Johnson, K. L., Macura, S., Juranic, N. & Kumar, R. (1997) Zinc binding properties of the DNA binding domain of the 1,25-dihydroxyvitamin d3 receptor. *Biochemistry* **36**, 10482-10491.

[133] Rice, L. M., Brennwald, P. & Brunger, A. T. (1997) Formation of a yeast snare complex is accompanied by significant structural changes. *FEBS Lett* **415**, 49-55.

[134] Permyakov, S. E., Millett, I. S., Doniach, S., Permyakov, E. A. & Uversky, V. N. (2003) Natively unfolded c-terminal domain of caldesmon remains substantially unstructured after the effective binding to calmodulin. *Proteins* **53**, 855-862.

[135] Parsegian, V. A., Rand, R. P., Fuller, N. L. & Rau, D. C. (1986) Osmotic stress for the direct measurement of intermolecular forces. *Methods Enzymol* **127**, 400-416.

[136] Parsegian, V. A., Rand, R. P. & Rau, D. C. (1995) Macromolecules and water: Probing with osmotic stress. *Methods Enzymol* **259**, 43-94.

[137] Parsegian, V. A., Rand, R. P. & Rau, D. C. (2000) Osmotic stress, crowding, preferential hydration, and binding: A comparison of perspectives. *Proc Natl Acad Sci U S A* **97**, 3987-3992.

[138] Fried, M. G., Stickle, D. F., Smirnakis, K. V., Adams, C., MacDonald, D. & Lu, P. (2002) Role of hydration in the binding of lac repressor to DNA. *J Biol Chem* **277**, 50676-50682.

[139] Fuxreiter, M., Simon, I., Friedrich, P. & Tompa, P. (2004) Preformed structural elements feature in partner recognition by intrinsically unstructured proteins. *J Mol Biol* **338**, 1015-1026.

[140] Bienkiewicz, E. A., Adkins, J. N. & Lumb, K. J. (2002) Functional consequences of preorganized helical structure in the intrinsically disordered cell-cycle inhibitor p27(kip1). *Biochemistry* **41**, 752-759.

[141] Lacy, E. R., Filippov, I., Lewis, W. S., Otieno, S., Xiao, L., Weiss, S., Hengst, L. & Kriwacki, R. W. (2004) P27 binds cyclin-cdk complexes through a sequential mechanism involving binding-induced protein folding. *Nat Struct Mol Biol* **11**,358-364.

[142] Lee, H., Mok, K. H., Muhandiram, R., Park, K. H., Suk, J. E., Kim, D. H., Chang, J., Sung, Y. C., Choi, K. Y. & Han, K. H. (2000) Local structural elements in the mostly unstructured transcriptional activation domain of human p53. *J Biol Chem* **275**, 29426-29432.

[143] Fletcher, C. M. & Wagner, G. (1998) The interaction of eif4e with 4e-bp1 is an induced fit to a completely disordered protein. *Protein Sci* **7**, 1639-1642.

[144] Mader, S., Lee, H., Pause, A. & Sonenberg, N. (1995) The translation initiation factor eif-4e binds to a common motif shared by the translation factor eif-4 gamma and the translational repressors 4e-binding proteins. *Mol Cell Biol* **15**, 4990-4997.

[145] Kissinger, C. R., Parge, H. E., Knighton, D. R., Lewis, C. T., Pelletier, L. A., Tempczyk, A., Kalish, V. J., Tucker, K. D., Showalter, R. E., Moomaw, E. W. & et al. (1995) Crystal structures of human calcineurin and the human fkbp12-fk506-calcineurin complex. *Nature* **378**, 641-644.

[146] Callaghan, A. J., Aurikko, J. P., Ilag, L. L., Gunter Grossmann, J., Chandran, V., Kuhnel, K., Poljak, L., Carpousis, A. J., Robinson, C. V., Symmons, M. F. & Luisi, B. F. (2004) Studies of the rna degradosome-organizing domain of the escherichia coli ribonuclease rnase e. *J Mol Biol* **340**, 965-979.

[147] Mohan, A. (2006) MoRFs: A dataset of molecular recognition features. Master of Science. The School of Informatics, Indiana University, Indianapolis. pp. 59

[148] Kelly, J. W. (1998) The alternative conformations of amyloidogenic proteins and their multi-step assembly pathways. *Curr Opin Struct Biol* **8**, 101-106.

[149] Dobson, C. M. (1999) Protein misfolding, evolution and disease. *Trends Biochem Sci* **24**, 329-332.

[150] Bellotti, V., Mangione, P. & Stoppini, M. (1999) Biological activity and pathological implications of misfolded proteins. *Cell Mol Life Sci* **55**, 977-991.

[151] Uversky, V. N., Talapatra, A., Gillespie, J. R. & Fink, A. L. (1999) Protein deposits as the molecular basis of amyloidosis. I. Systemic amyloidoses. *Med. Sci. Monitor* **5**, 1001-1012.

[152] Uversky, V. N., Talapatra, A., Gillespie, J. R. & Fink, A. L. (1999) Protein deposits as the molecular basis of amyloidosis. Ii. Localized amyloidosis and neurodegenerative disordres. *Med. Sci. Monitor* **5**, 1238-1254.

[153] Rochet, J. C. & Lansbury, P. T., Jr. (2000) Amyloid fibrillogenesis: Themes and variations. *Curr Opin Struct Biol* **10**, 60-68.

[154] Uversky, V. N. & Fink, A. L. (2004) Conformational constraints for amyloid fibrillation: The importance of being unfolded. *Biochim Biophys Acta* **1698**, 131-153.
[155] Uversky, V. N. & Fink, A. L. (2005) Pathways to amyloid fibril formation: Partially folded intermediates in fibrillation of unfolded proteins. in *Amyloid proteins: The beta pleated sheet conformation and disease* (Sipe, J. D., Ed.) pp 247-265, Wiley-VCH, Verlag GmbH & Co. KGaA, Weinheim, Germany.
[156] Glenner, G. G. & Wong, C. W. (1984) Alzheimer's disease and down's syndrome: Sharing of a unique cerebrovascular amyloid fibril protein. *Biochem Biophys Res Commun* **122**, 1131-1135.
[157] Masters, C. L., Multhaup, G., Simms, G., Pottgiesser, J., Martins, R. N. & Beyreuther, K. (1985) Neuronal origin of a cerebral amyloid: Neurofibrillary tangles of alzheimer's disease contain the same protein as the amyloid of plaque cores and blood vessels. *Embo J* **4**, 2757-2763.
[158] Lee, V. M., Balin, B. J., Otvos, L., Jr. & Trojanowski, J. Q. (1991) A68: A major subunit of paired helical filaments and derivatized forms of normal tau. *Science* **251**, 675-678.
[159] Ueda, K., Fukushima, H., Masliah, E., Xia, Y., Iwai, A., Yoshimoto, M., Otero, D. A., Kondo, J., Ihara, Y. & Saitoh, T. (1993) Molecular cloning of cdna encoding an unrecognized component of amyloid in alzheimer disease. *Proc Natl Acad Sci U S A* **90**, 11282-11286.
[160] Wisniewski, K. E., Dalton, A. J., McLachlan, C., Wen, G. Y. & Wisniewski, H. M. (1985) Alzheimer's disease in down's syndrome: Clinicopathologic studies. *Neurology* **35**, 957-961.
[161] Dev, K. K., Hofele, K., Barbieri, S., Buchman, V. L. & van der Putten, H. (2003) Part ii: Alpha-synuclein and its molecular pathophysiological role in neurodegenerative disease. *Neuropharmacology* **45**, 14-44.
[162] Prusiner, S. B. (2001) Shattuck lecture--neurodegenerative diseases and prions. *N Engl J Med* **344**, 1516-1526.
[163] Zoghbi, H. Y. & Orr, H. T. (1999) Polyglutamine diseases: Protein cleavage and aggregation. *Curr Opin Neurobiol* **9,** 566-570.
[164] Uversky, V. N. (2003) A protein-chameleon: Conformational plasticity of alpha-synuclein, a disordered protein involved in neurodegenerative disorders. *J Biomol Struct Dyn* **21**, 211-234.
[165] Okazawa, H. (2003) Polyglutamine diseases: A transcription disorder? *Cell Mol Life Sci* **60**, 1427-1439.
[166] Cummings, C. J. & Zoghbi, H. Y. (2000) Fourteen and counting: Unraveling trinucleotide repeat diseases. *Hum Mol Genet* **9,** 909-916.
[167] Gusella, J. F. & MacDonald, M. E. (2000) Molecular genetics: Unmasking polyglutamine triggers in neurodegenerative disease. *Nat Rev Neurosci* **1**, 109-115.
[168] Orr, H. T. (2001) Beyond the qs in the polyglutamine diseases. *Genes Dev* **15**, 925-932.
[169] Fischbeck, K. H. (2001) Polyglutamine expansion neurodegenerative disease. *Brain Res Bull* **56**, 161-163.
[170] Adkins, J. N. & Lumb, K. J. (2002) Intrinsic structural disorder and sequence features of the cell cycle inhibitor p57kip2. *Proteins* **46**, 1-7.
[171] Chang, B. S., Minn, A. J., Muchmore, S. W., Fesik, S. W. & Thompson, C. B. (1997) Identification of a novel regulatory domain in bcl-x(l) and bcl-2. *Embo J* **16**, 968-977.

[172] Campbell, K. M., Terrell, A. R., Laybourn, P. J. & Lumb, K. J. (2000) Intrinsic structural disorder of the c-terminal activation domain from the bzip transcription factor fos. *Biochemistry* **39**, 2708-2713.

[173] Sunde, M., McGrath, K. C., Young, L., Matthews, J. M., Chua, E. L., Mackay, J. P. & Death, A. K. (2004) Tc-1 is a novel tumorigenic and natively disordered protein associated with thyroid cancer. *Cancer Res* **64**, 2766-2773.

[174] Linding, R., Jensen, L. J., Diella, F., Bork, P., Gibson, T. J. & Russell, R. B. (2003) Protein disorder prediction: Implications for structural proteomics. *Structure (Camb)* **11**, 1453-1459.

[175] Ward, J. J., Sodhi, J. S., McGuffin, L. J., Buxton, B. F. & Jones, D. T. (2004) Prediction and functional analysis of native disorder in proteins from the three kingdoms of life. *J Mol Biol* **337**, 635-645.

[176] Prilusky, J., Felder, C. E., Zeev-Ben-Mordehai, T., Rydberg, E. H., Man, O., Beckmann, J. S., Silman, I. & Sussman, J. L. (2005) Foldindex: A simple tool to predict whether a given protein sequence is intrinsically unfolded. *Bioinformatics* **21**, 3435-3438.

[177] Linding, R., Russell, R. B., Neduva, V. & Gibson, T. J. (2003) Globplot: Exploring protein sequences for globularity and disorder. *Nucleic Acids Res* **31**, 3701-3708.

[178] Dosztanyi, Z., Csizmok, V., Tompa, P. & Simon, I. (2005) Iupred: Web server for the prediction of intrinsically unstructured regions of proteins based on estimated energy content. *Bioinformatics* **21**, 3433-3434.

[179] Liu, J. & Rost, B. (2003) Norsp: Predictions of long regions without regular secondary structure. *Nucleic Acids Res* **31**, 3833-3835.

[180] Romero, P., Obradovic, Z., Kissinger, C. R., Villafranca, J. E., Garner, E., Guilliot, S. & Dunker, A. K. (1998) Thousands of proteins likely to have long disordered regions. *Pac Symp Biocomput*, 437-448.

[181] Coeytaux, K. & Poupon, A. (2005) Prediction of unfolded segments in a protein sequence based on amino acid composition. *Bioinformatics* **21**, 1891-1900.

[182] Yang, Z. R., Thomson, R., McNeil, P. & Esnouf, R. M. (2005) Ronn: The bio-basis function neural network technique applied to the detection of natively disordered regions in proteins. *Bioinformatics* **21**, 3369-3376.

[182] Wootton, J. C. (1994) Non-globular domains in protein sequences: Automated segmentation using complexity measures. *Comput Chem* **18**, 269-285.

[183] Ekman, D., Light, S., Bjorklund, A. K., and Elofsson, A. (2006) What properties characterize the hub proteins of the protein-protein interaction network of Saccharomyces cerevisiae? *Genome Biol* **7**, R45.

[184] Patil, A., and Nakamura, H. (2006) Disordered domains and high surface charge confer hubs with the ability to interact with multiple proteins in interaction networks. *FEBS Lett* **580**, 2041-5.

[185] Dosztányi, Z., Chen, J., Dunker, A. K., Simon, I., and Tompa, P. (2006) Disorder and sequence repeats in hib proteins and their implications for network evolution. *Journal of Proteome Research* **5,** 2985-2995.

[186] Haynes, C., Oldfield, C. J., Ji, F., Klitgord, N., Cusick, M. E., Radivojac, P., Uversky, V. N., Vidal, M., and Iakoucheva, L. M. (2006) Intrinsic disorder is a common feature of hub proteins from four eukaryotic interactomes. *PLoS Comput Biol* **2**, e100.

[187] Bustos, D. M., and Iglesias, A. A. (2006) Intrinsic disorder is a key characteristic in partners that bind 14-3-3 proteins. *Proteins* **63**, 35-42.

[188] Radivojac, P., Vucetic, S., O'Connor, T. R., Uversky, V. N., Obradovic, Z., and Dunker, A. K. (2006) Calmodulin signaling: analysis and prediction of a disorder-dependent molecular recognition. *Proteins* **63**, 398-410.

[189] Mohan, A., Oldfield, C. J., Radivojac, P., Vacic, V., Cortese, M. S., Dunker, A. K., and Uversky, V. N. (2006) Analysis of Molecular Recognition Features (MoRFs). *J Mol Biol* **362**, 1043-59.

In: Protein Folding-Misfolding…
Editors: J.P. Zbilut and T. Scheibel, pp. 93-116

ISBN: 1-60021-417-7


*Chapter 4*

# PROTEIN PRODUCTION IN THE TEST TUBE: *IN VITRO* TRANSCRIPTION / TRANSLATION SYSTEMS IN PROTEIN EXPRESSION AND FOLDING

***Martin Haslbeck*, Stephan Frey and Johannes Buchner***
Technische Universität München, Department Chemie, Lehrstuhl Biotechnologie
Lichtenbergerst. 4, D-85747 Garching, Germany

## Abstract

After sequencing of a substantial set of genomes is completed, one of the key problems of today's structural genomics and proteomics projects is high-throughput expression and purification of a large number of target proteins. The efficient production of many gene products is required in therapeutic research or for the decoding of the function of all proteins of one organisms. Traditionally, the production of a recombinant protein requires a cloning step of the target gene into an expression vector followed by its expression in bacteria, yeast or mammalian and insect cells. However, in many cases the recombinant proteins are not produced in substantial yields and in an active form. Thus, the development of efficient expression and purification systems is a major goal. While conventional microbial *in vivo* expression systems seem rather ill-suited for high throughput technologies, cell-free systems might be the tools of choice for the fast expression of proteins in parallel and for screening of a wide range of expression conditions. In addition, *in vitro* translation systems turned out to be very useful to study the influence of molecular chaperones on protein expression and folding.

**Keywords:** cell-free protein synthesis, protein expression, molecular chaperones, *in vitro* translation / transcription

* E-mail address: martin.haslbeck@ch.tum.de; Phone: +49-89-289-1 33 67; Fax: +49-89-289-1 33 45 (Corresponding author)

## Abbreviations

AcP, acetyl phosphate; ARSes, aminoacyl-tRNA synthetases (ribosome bound); CECF, continuous exchange cell free; CFCF, continuous flow cell free; CK, creatine kinase; CP, creatine phosphate; DNA, desoxyribonucleic acid; DTT, dithiotreitol; EF, elongation factor; FAD, Flavin adenin dinukleosid; GFP, green fluorescent protein; Hsp, heat shock protein; IF, initiations factor; kDa, kilo Dalton; ME, mercaptoethanol; MIA, Melanoma Inhibitory Activity; mRNA, messanger RNA; NAPPA, nucleic acid progarmmable protein array; NMR, nuclear magnetic resonance; NTPs, nucleosid triphosphates; PCR, polymerase chain reaction; PEP, phosphoenol pyruvate; $P_i$, orthophosphate; PK, pyruvate kinase; $PP_i$, pyrophosphate; RF, termination factor; RNA, ribonucleic acid; rpm, rounds per minute; RTS, rapid translation system; SGO,Structural Genomics Organization; sHsp, small Hsp; tRNA, transfer RNA

## 1. Introduction

The genome sequences of a growing variety of organisms are now available. Having identified the set of proteins that determine an organism, the analysis of their individual structures, function and interactions in a systems approach is the task for the future (Yokoyama, 2003; Spirin, 2004; Vitkup et al., 2001; Kim, 2000). At the moment, more than 25 projects have joined the International Structural Genomics Organization (SGO, http://ww.isgo.org) and projects on the proteoms of several organisms including man (HUPO, http://www.hupo.org) have been started. In the years ahead, the expression, purification and characterization of thousands of proteins and the determination of more than 10 000 representative protein structures is planned (Vitkup et al., 2001; Yokoyama, 2003).

In general, recombinant protein production processes consist of three modules, gene cloning, expression and protein purification. For systems biology, these modules should allow the parallel production of hundreds of target proteins. Although many of the organisms studied are eukaryotes, recombinant expression in prokaryotes, mainly in *E. coli*, is still the first choice (Yokoyama, 2003; Hurley et al., 2002; Christendat et al., 2000a; Yee et al., 2003). Automated, parallel cloning to produce bacterial expression plasmids is meanwhile possible in a high throughput format (Dieckman et al., 2002; Dieckman et al., 2006). But still these cloning strategies need some time due to the growth period while selecting positive clones, and they are rather expensive. In addition to these limitations, *E. coli* is often not suited to produce the desired target proteins in sufficient amounts. Especially eukaryotic, large or disulfide-bridged proteins sometimes cause tremendous complications when expressed in *E. coli*. Also the biosynthetic isotope labeling of target proteins for structural studies like NMR or crystallography, is expensive. To overcome theses problems *in vitro* protein synthesis systems may become important tools for high throughput protein expression (Hoffmann et al., 2004; Yokoyama, 2003; Spirin, 2004; Swartz et al., 2004; Nakano et al., 2004; Jackson et al., 2004).

**Table 1. Comparison of selected *in vitro* transcription / translation systems**

| **System** | Crude bacterial cell free translation system (Matthaei and Nirenberg, 1961). | Crude bacterial coupled transcription / translation system (DeVries and Zubay, 1967). | Crude eukaryotic combined transcription / translation system (Baranov and Spirin, 1993). | Pure DNA-directed cell-free transcription / translation system (Kung et al., 1977). |
|---|---|---|---|---|
| **Source** | supernatant fraction S-30 of *E.coli* | supernatant fraction S-30 of *E.coli* | supernatant fractions S-30 or S-100 of wheat germ or rabbit reticulocytes | *E.coli* purified components |
| **Pre-treatment** | DNase; preincubated with amino acids, PEP and PK at 37°C for 80 min and then dialyzed | preincubated to run of endogenous DNA and RNA | treated with micrococcal $Ca^{2+}$ dependent RNase (in case of reticulocyte lysate) | |
| **Source components:** | | | | |
| | - ribosomes<br>- tRNAs<br>- ARSes<br>- translation factors<br>- natural mRNAs | - RNA polymerase<br>- ribosomes<br>- tRNAs<br>- ARSes<br>- translation factors | - ribosomes<br>- tRNAs<br>- ARSes<br>- translation factors | - ribosomes<br>- formyltetrahydrofolate Met-$tRNA_f$ transformylase<br>- initiation factors: IF1, IF2, IF3<br>- elongation factors: EF-Tu, EF-Ts, EF-G<br>- termination factors: RF1, RF2, RF3, RRF |
| **Added components** | | | | |
| | - synthetic polyribo-nucleotides<br>- amino acids | - amino acids | - DNA (plasmid with SP6 or T7 promotor)<br>- bacteriophage SP6 or T7 polymerase<br>- amino acids | - DNA (target gene)<br>- total tRNA<br>- RNA polymerase<br>- 20 individual ARSes<br>- amino acids |
| **Nucleotides** | ATP / GTP | ATP, GTP, UTP, CTP | ATP, GTP, UTP, CTP | ATP, GTP, UTP, CTP |
| **Regenerating system:** | PEP + PK | PEP + PK, or CP + CK, or AcP | CP + CK | PEP + PK |
| **SH-compound** | DTT or ME | DTT or ME | DTT | DTT |
| **Ions** | $Mg^{2+}$, $K^+$, $NH_4^+$, in optimal concentrations | $Mg^{2+}$, $K^+$, $NH_4^+$, in optimal concentrations | $Mg^{2+}$, $K^+$ and polyamines in concentrations optimal for translation; $Ca^{2+}$ for wheat germ lysate | $Mg^{2+}$, $K^+$, $NH_4^+$, in optimal concentrations |
| **Further compounds** | | formyltetrahydrofolate or cAMP | hemin, gluthatione and cAMP may be useful for the reticulocyte system | methenyltetrahydro-folate and spermidine |

Coupled *in vitro* transcription / translation systems allow the specific manipulation of the reaction conditions during protein biosynthesis (Fig. 1; Spirin et al., 1988; Matthaei and Nirenberg, 1961; DeVries and Zubay, 1967). In these open systems, the addition and depletion of components is possible (Table 1). In contrast to *in vivo* expression systems, they allow the expression of toxic, regulatory or disulfide-bonded proteins, as well as the incorporation of unnatural amino acids.

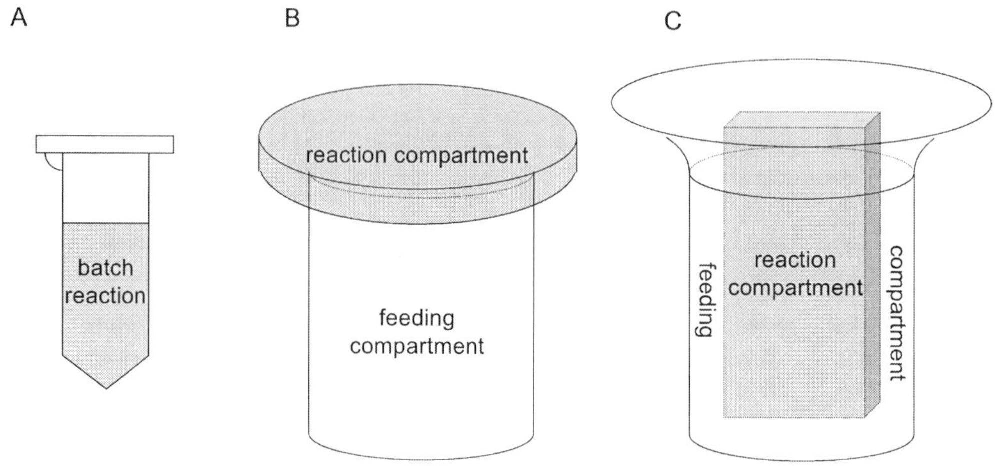


**A:** Typical batch system, where a small volume of reaction mixture (Table 1) is simply filled in a test tube.
**B:** Typical CECF system where the reaction compartment is separated from a feeding compartment via a semipermeable membrane. Commonly the reaction compartment is ~ 10 times smaller than the feeding compartment. Exchange of components takes place passively driven by diffusion and might be enhanced by stirring of both compartments.
**C:** Typical CEFC system, where the reaction compartment consists of a *dialysis bag*, hung in a feeding compartment. Compared to systems described in **B**, the passive exchange of components is often favored in this type of setup and the setup can be scaled up easily. On the other hand addition of components and stirring of the reaction compartment is more complicated.

Figure 1. Comparison of batch and continuous *in vitro* transcription / translation systems

More than 40 years ago, first experiments to synthesize proteins *in vitro* in cell-free systems were already performed (Nirenberg and Matthaei, 1961). In those ancient expression systems only a short expression period and, as a result, only low amounts of product could be achieved. Today, these systems are mainly used for the synthesis of radio-labeled proteins. Spirin and his coworkers were finally able to establish a cell-free expression system with a longer period of protein synthesis (Spirin et al., 1988; Baranov et al., 1989). The underlying idea of this system is the use of two compartments, split by a semi-permeable membrane. This allows the continuous delivery of energy and substrates from the feeding compartment to the reaction compartment of the transcription / translation system (Fig. 1). As a result, the expression period and the resulting amounts of expressed protein increase dramatically. Nevertheless, unspecific aggregation of the target protein is still one of the biggest problems during recombinant expression of proteins in *in vitro* transcription / translation systems, as well as in the cell (Christendat et al., 2000b; Christendat et al., 2000a). Thus, *in vitro* transcription / translation systems are also suitable to study protein folding and the influences of different additives, allowing to screen a large set of parameters in parallel for effective optimization of protein production. In addition, the basic mechanisms of molecular chaperones and folding helper proteins can be studied and exploited easily in these test tube systems (Yin and Swartz, 2004; Haslbeck et al., 2005b; Stemp et al., 2005; Hofweber et al., 2005).

## 2. *In vitro* Transcription / Translation Systems

A) Batch systems

Strikingly, a fundamental portion of our understanding of protein synthesis including the discovery of mRNA and the genetic code as well as the function of ribosomes was originally obtained utilizing bacterial extracts and *in vitro* translation reactions (Spirin, 2004; Nirenberg, 2004; Spirin, 2002a). Since the first experiments in the 1950s, protein synthesis *in vitro* has been improved continuously. After the elementary demonstration that protein synthesis *per se* does not require the integrity of the cell and may be continued in cell extracts (Borosook et al., 1950; Siekevitz and Zamecnik, 1981; Spirin, 2002a), a further breakthrough was the discovery of ribonucleoprotein particles with protein synthesizing ability, i.e. ribosomes (Tissieres and Watson, 1962; Tissieres and Watson, 1958; Chao and Schachman, 1956; Littlefield et al., 1955; Spirin, 2002a). A cell-free, ribosomal system for protein synthesis based on cytoplasmic extracts of animal cells was already established in this decade by Zamecnik and co workers (Keller and Zamecnik, 1956). A real revolution in the field was the reprogramming of the system with exogenous mRNA (Nirenberg and Matthaei, 1961). This classical system (Table 1) still is the basis for several commercial and most self-made *in vitro* translation systems. A further breakthrough was the coupling of transcription and translation in the *in vitro* systems (DeVries and Zubay, 1967; Zubay, 1973). Here bacterial crude extracts were used in which endogenous RNA and DNA were degraded by cellular nucleases (Table 1). The addition of exogenous DNA, such as a plasmid or a synthetic DNA fragment, results in the synthesis of the corresponding mRNA by the RNA polymerase present in the extract or added directly.

In the seventies, the combination of eukaryotic extracts (commonly rabbit reticulocyte lysate or wheat germ extract) with bacterial or viral polymerases defined a new sort of systems (Table 1; Roberts and Paterson, 1973; Roberts et al., 1975; Stueber et al., 1984; Bujard et al., 1987). Eukaryotic extracts were prepared from the cytosolic fraction after cell disruption, leading to the lack of nuclear localized endogenous RNA polymerase activity. The addition of exogenous *E. coli* RNA polymerase resulted in highly effective systems (Spirin, 2002a). The addition of bacteriophage T7 or SP6 RNA polymerase proofed to be even more efficient (Baranov and Spirin, 1993; Baranov et al., 1989; Spirin, 2002a; Craig et al., 1992; Kigawa et al., 1999). Furthermore, it offers the possibility to limit the transcription to DNA containing the specific promoter regions of the respective polymerases. At present, these systems represent a significant portion of the commercial market. However, while ribosomes start to translate the nascent mRNA prior to completion of RNA synthesis in entirely bacterial systems (e.g. *coupled transcription / translation systems*), in *combined transcription / translation systems* mRNA is synthesized mainly in advance since the bacteriophage RNA polymerases work much faster than the eukaryotic translation system (Table 1).

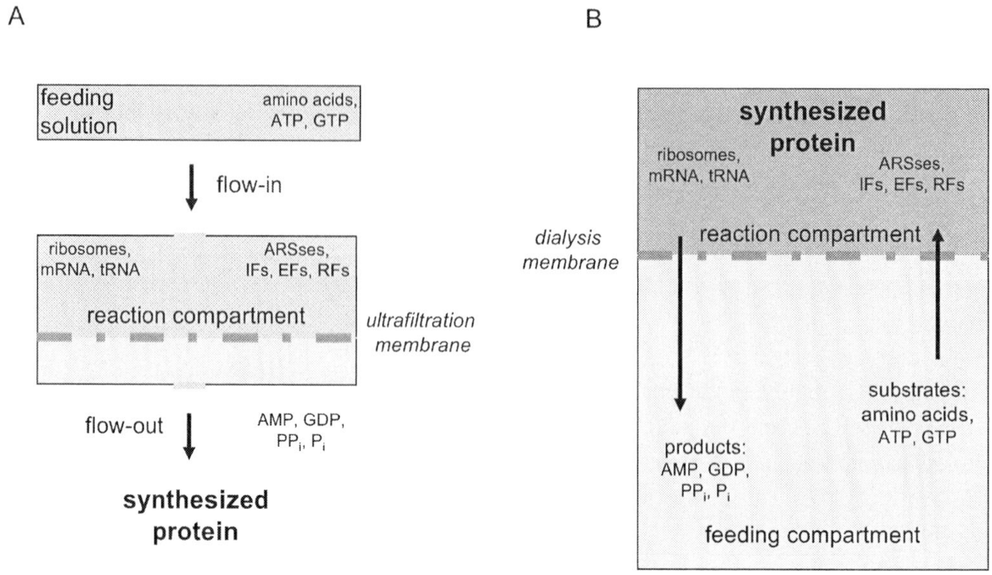


A: The setup of a CFCF system, where the feeding solution is actively pumped through the reaction compartment and the synthesized product flows actively out of the reaction mixture (Spirin, 2002b) is shown.

B: The passive flow of the components in a CECF system is indicated. Stirring of both compartments efficiently favors the exchange of components as achieved in the RTS.

Figure 2. Comparison of the principal setup and flow scheme of a CECF and CFCF system

All the systems described so far, are classical *batch systems*, where all components are mixed together in a small volume in a test tube (Fig. 1A). They all show only low yields of protein synthesis because of the short performance time of the system due to inhibiting reaction products and limited energy supply (Kim and Swartz, 1999; Spirin, 2004). Thus, most of the commercial systems are only used for analytical or screening purposes and are not suitable for protein synthesis on a preparative scale. A typical bacterial batch system for example is only active for 10-30 min at 37°C and a batch system based on rabbit reticulocyte lysate or wheat germ extract might be productive approximately for 1 h (Spirin, 2002a).

B) Continuous systems

To overcome this principal problem of short performance times, *continuous exchange cell-free systems* (CECF) were established (Spirin et al., 1988; Spirin, 2002a; Spirin, 2004). They consist of two compartments separated by a porous barrier, e.g. a dialysis membrane (Fig. 1A/B and Fig. 2). One compartment contains all the reaction components, while the second compartment supplies the reaction with consumable substrates (amino acids, NTPs and energy compounds) and allows the removal of reaction products (inorganic phosphate and nucleoside monophosphates). The two compartments might be separated by a flat dialysis membrane (Fig. 1B) or a membrane bag might be hung into a feeding compartment (Fig. 1C). In consequence, the performance time of the transcription / translation systems was prolonged to 14-20 h and yields of 1-6 mg /ml were achieved for several target proteins (Kigawa et al., 1999; Kigawa et al., 2004; Madin et al., 2000; Hoffmann et al., 2004).

In addition to CECF systems, *continuous flow cell-free systems* (CFCF) were described (Fig. 2). Here, a feeding solution containing the consumable substrates is continuously pumped into the reaction, while the same volume containing the synthesized product is flowing out of the system via a permeable membrane (Baranov et al., 1989; Spirin, 2002a). The rate of flow through is typically between 0.5 and 3 volumes of the reaction compartment per hour. These systems allow constant protein synthesis over 20 to 100 hours using bacterial as well as in eukaryotic lysates (Baranov and Spirin, 1993).

Interestingly, no significant leakage of components of the protein-synthesizing machinery (tRNAs, translation factors, aminoacyl-tRNA synthetases) through 50-300 kDa cut-off permeable membranes was observed, while the synthesized product of comparable molecular mass was removed from the reaction (Spirin, 2002a). This could be explained by the formation of large dynamic complexes in the process of transcription and translation leaving no significant amounts of free functional components.

When CECF and CFCF systems are compared with each other (Fig. 2), one advantage of the CEFC is obviously its simplicity and lower costs. On the other hand, CFCF is able to synthesize target proteins with an exceptional purity of up to 85% and with a tremendously extended performance time (Spirin, 2002b). In addition, the up-scaling of CFCF systems is much easier because of the more efficient exchange of compounds forced by the applied flow. This regulated flow allows also the direct online control and therefore programming of the reaction composition and the out-flowing product.

C) Systems assembled of purified components

In all cell-free protein synthesis systems described so far, the basis of the entire system was a crude cell extract of some sort. Since the isolation and purification of single proteins made extensive progress in the last decades, it is meanwhile possible to reconstitute cell-free protein synthesizing systems from well-characterized, highly purified components (Table 1; Kung et al., 1977; Zarucki-Schulz et al., 1979; Green et al., 1985; Shimizu et al., 2001). Of course these systems are extremely expensive and might not be the first choice for high throughput protein synthesis. However, the defined composition enables studies on the function and influences of single components as well as the use of engineered proteins and other compounds as part of the synthesizing machinery

## 3. Rapid Translation System

The commercially available Rapid Translation System (RTS; Roche Diagnostics, Mannheim, Germany) is a convenient, preparative-scale, cell-free protein expression which incorporates the CECF technology (Fig. 2; Spirin et al., 1988; Hoffmann et al., 2004; Martin et al., 2001). The technology consists of a collection of devices for protein production in different scales, ranging from 50 µl to 10 ml systems. The central device of this technology consists of a 1-ml reaction chamber and a 10-ml feeding chamber separated by a 10-kDa semi-permeable dialysis membrane (Fig 1B). T7 RNA polymerase catalyzed transcription and *E. coli* lysate driven translation take place simultaneously in the reaction compartment. Substrates and energy components essential for sustained protein expression are continuously supplied from the feeding compartment into the reaction compartment through the semi-permeable dialysis membrane. As the reaction proceeds, inhibitory waste products and reaction by-products diffuse down the concentration gradient into the feeding compartment. Thus, accumulation of inhibitory waste components and depletion of substrates is avoided. The exchange of components is enhanced by mixing the compartments continuously with free rotating magnetic stirring bars. Commonly, protein is expressed for up to 24 hours, yielding up to 5 mg of protein per ml (Hoffmann et al., 2004).

The workflow from a gene of interest to its preparative synthesis in the RTS might start with any sort of DNA carrying a T7 promotor. An advantage over cellular expression is that this DNA construct must not necessarily be a plasmid, a linear expression cassette generated by PCR is also sufficient. In a two step PCR reaction, the T7 promotor region and the terminator is fused to the gene of interest. Hence, cloning can be avoided and protein production is speeded up tremendously. The whole process of two PCR steps followed by *in vitro* expression is feasible in less than 16 h. In addition, the RTS offers the possibility to optimize the sequence in terms of codon usage or minimizing secondary structures of the mRNA in the promoter region during these initial PCR steps (www.proteoexpert.com; (Voges et al., 2004; Paulus et al., 2004)). While it is rather complicated and time-consuming to perform these optimizations for *in vivo* expression, it is advisable to optimize *in vitro* transcription / translation to force target gene expression. For screening purposes, a subset of generated linear expression constructs can be tested in a 50 to 100 µl batch or CECF reaction (RTS 100) in parallel. All expressing constructs can be used directly in larger scale CECF reactions. Nevertheless, it should be kept in mind that for large scale expression the generation of a circular expression template is easier and cheaper to handle. The generation of circular templates can be achieved by inserting the expression cassettes into PCR cloning vectors (such as TOPO, Invitrogen, Karlsruhe, Germany; BD In-Fusion, BD Biosciences, Erembodegem, Belgien; pIVEX, Roche Diagnostics, Mannheim, Germany).

Optimization of the expression of the desired target protein can be achieved already in small-scale reactions. Classical parameters to test are expression temperature and reaction time. For example, MIA (Melanoma Inhibitory Activity) protein, a secreted protein from human melanoma cell lines, is not correctly folded and is inactive after synthesis in bacteria (Guba et al., 2000; Bosserhoff, 2003). Expression screening in the RTS showed that the highest amount of synthesised target protein was achieved at 30°C, whereas the percentage of correctly folded protein was higher at 25°C (Bosserhoff, 2003). Furthermore, the co-expression or addition of co-factors, binding partners or protein subunits can be screened easily to accomplish efficient target protein synthesis and folding. In the case of Kox1, a member of the KRAB zinc finger protein family, the addition or co-expression of its interaction partner TIF1ß was necessary to gain properly folded protein (Lorenz and Thiesen, 2003). In another study, two fragments of the *E. coli* protein MalE, an exported periplasmic receptor for high-affinity transport of maltodextrins, were expressed separately but assembled correctly in the RTS (Betton and Hofnung, 1994; Betton, 2003). In this example, especially proteolytic degradation of the target protein in the RTS was substantially lower than *in vivo*.

A further example is D-amino acid oxidase (DAO) from *T. variabilis*, a FAD-containig oxidase (D'Aniello et al., 1993). The expression of the protein in *E. coli* interfered with the growth of the *E. coli* cells by disturbance of the cell wall metabolism and/or toxic effects through enzymatically produced $H_2O_2$. In the RTS, the protein could be expressed in soluble and functional active form, when the essential cofactor FAD was added in sufficient amount (Lehmann and Thiesen, 2003).

Another very interesting feature in terms of optimization the open reaction is the use of the RTS to specifically label the target protein. In the last decade, it became clear that the set of 20 amino acids specified by the universal genetic code can be expanded (Kobayashi et al., 2005; Leconte et al., 2005; Hyun et al., 2003). Canonical amino acids attached to cognate tRNAs can be modified or designed amino acid versions can be loaded onto desired tRNAs (Hyun et al., 2003). Specific protein labelling with modified amino acids is used in a wide variety of experimental setups mainly with the purpose of structure determination. The incorporation of electron-rich selenium-containing non-canonical counterparts of Trp and Met into proteins via *in vitro* translation represents a useful route for solving the phase problem in protein X-ray crystallography (Kigawa et al., 2002; Johansson et al., 2005). Because of the small reaction volume used for *in vitro* translation systems, the amounts of labelled amino acids needed and the costs involved decline dramatically compared to *in vivo* labelling. In similar approaches, the efficient labelling of proteins with $^{15}N$, $^{13}C$ and $^{2}H$ for subsequent NMR studies was demonstrated in the RTS (Kigawa et al., 1999; Kigawa et al., 1995; Scott et al., 2004; Staunton et al., 2006). Here again *in vitro* synthesis has some important advances compared to *in vivo* labelling. A major problem results from metabolic reactions which alter the amino acids once they are taken up by *E. coli.* Aspartate can be converted into glutamate, asparagines and lysine. The addition of labelled aspartate might therefore result in the inadvertent labelling of several amino acids (McIntosh and Dahlquist, 1990; Zhao et al., 2002). Similar metabolic interconversion pathways exist for many other amino acids as well. Furthermore, due to ineffective uptake and incorporation significant amounts of the expensive labelled amino acids are wasted. In CECF systems like the RTS, the labelled amino acids are restricted to the reaction compartment and they outcompete completely the unlabelled ones. Thus, the expressed protein is labelled with very high efficiency and exclusively because the target protein is the only protein synthesized. In addition, due to the absence of metabolism, no scrambling of the label occurs. In cases where the expression yield is high enough (~5 mg/ml) the structure determination of the newly synthesized labelled protein by NMR is possible directly in the reaction-mixture without further purification of the target protein (Yee et al., 2003; Guignard et al., 2002; Ozawa et al., 2005). These observations demonstrate that cell-free protein expression for structure determination of proteins has unique advantages over cellular expression systems regarding labelling efficiency, speed and costs.

As a further advanced application, it has been demonstrated that the RTS allows the formation of disulfide bonds, when the free sulfhydryl groups of endogenous lysate proteins are blocked by sulfhydryl-specific alkylating agents (Yin and Swartz, 2004). A modified version of the RTS (RTS *E. coli* disulfide Kit; Roche Diagnostics, Mannheim, Germany), where the *E. coli* lysate is treated with iodacetamide, shows no reducing activity and allows the efficient production of disulfide bridged proteins, without influencing the capacity of protein synthesis. Iodacetamide was chosen as the thiol-modifying agent since it efficiently blocks thiol-mediated catalysis without changing the ionic charge of the proteins. The further addition of oxidizing glutathione together with the addition of a disulfide isomerase (DsbC) enabled the expression of active murine urokinase (Yin and Swartz, 2004; Kim et al., 2003).

The biggest problem limiting the yields of *in vitro* transcription / translation systems is aggregation and thus, production of inactive target protein. Expression of recombinant proteins in *E. coli* cells often results in the aggregation of the produced target protein and this effect is also observed in cell-free transcription / translation reactions, especially when large mammalian proteins are produced or when the production rate and the end concentration of target protein are rather high. An opportunity to overcome this aggregation problem is to supplement the reaction with molecular chaperones.

## 4. Molecular Chaperones

Whereas many proteins can be refolded *in vitro* under optimized conditions with good yields, the situation in a living cell is less favorable. In particular, high protein concentrations and high temperature promote aggregation as an undesired side reaction competing with productive folding (Jaenicke and Seckler, 1997; Jaenicke, 1998; Walter and Buchner, 2002; Dill and Chan, 1997). To overcome this problem, a specialized family of folding helpers, the molecular chaperones, evolved (Walter and Buchner, 2002; Bukau et al., 2006; Hartl and Hayer-Hartl, 2002). They seem to represent a general solution for folding problems in protein production.

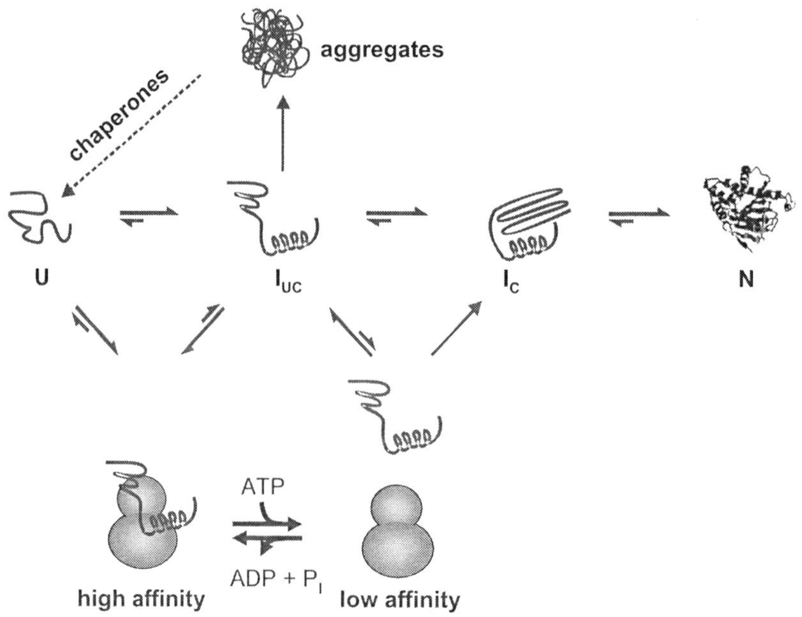


Protein biosynthesis as well as cellular stress results in the formation of unfolded polypeptides [U]. These molecules will fold via several intermediates [$I_{uc}/I_c$] with increasing amounts of structure until they reach the native, functional state [N]. Some intermediates [$I_{uc}$] may expose hydrophobic surfaces that render them susceptible for aggregation. Molecular chaperones interfere with the deleterious process of aggregation by binding to species [$I_{uc}$] and [U]. This association not only blocks the hydrophobic patches on the bound polypeptides but also decreases the concentration of aggregation-prone molecules thereby slowing down aggregation. In many cases, an ATP-mediated conformational change in the chaperone triggers the dissociation of the bound polypeptide. A fraction of the released molecules may fold into a committed state [$I_c$], which no longer requires the assistance of the chaperone, whereas the remaining uncommitted ($I_{uc}$) molecules rebind and participate in another chaperone cycle (Walter and Buchner, 2002).

Figure 3. A model for the principle of chaperone function

Under physiological and especially under stress conditions, molecular chaperones are involved in the maintenance of cellular processes, the mediation of stress tolerance and in suppression of protein aggregation in general. They fulfill their task by interacting with a broad range of unfolded proteins (Fig. 3). They recognize their substrates via exposed hydrophobic surfaces and not via specific peptides sequences or defined structural elements. This hydrophobic interaction enables molecular chaperones to distinguish between native, correctly folded and non-native proteins (see Fig. 3). Chaperones do not contribute structural information for folding, but prevent unwanted intermolecular interactions by binding to the folding polypeptide chain (Walter and Buchner, 2002; Hartl and Hayer-Hartl, 2002; Bukau et al., 2006). In contrast to the unspecific interactions of proteins during aggregation, chaperones form defined complexes with non-native proteins. The controlled release of the non-native proteins from the complex is normally accomplished by a change of the affinity of the chaperone for its substrate. According to the laws of thermodynamics, bound non-native protein stabilizes a high affinity state of the chaperone. Therefore, energy is needed to produce a state of low substrate affinity. Usually this energy is provided by the hydrolysis of ATP or the interaction with other chaperones (co-chaperones; see Fig. 3).

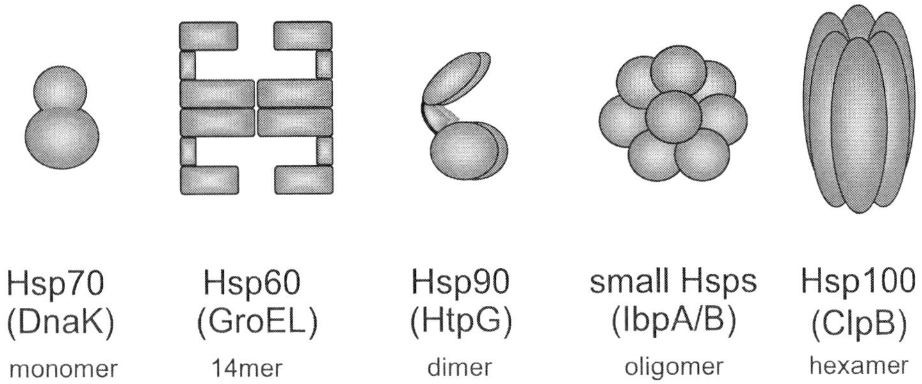


Under physiological and especially under stress conditions five different families of chaperones, named according to their molecular mass, are involved in the maintenance of cellular processes. The *E. coli* representatives of each family and the respective functional state of oligomerisation are indicated.

Figure 4. The five major families of chaperones

The family of molecular chaperones can be divided into several distinct groups (Fig. 4), considering sequence homologies and molecular masses: the small heat shock proteins (sHsps) with a molecular mass ranging from 10 to 40 kDa (Haslbeck, 2005); the 70 kDa Hsps, e. g. Hsp70, DnaK, and additionally their assisting Hsp40 proteins, e. g. DnaJ (Bukau et al., 2000); the 60 kDa Hsps, e. g. GroEL, and its assisting co-chaperone GroES (Horwich et al., 2006; Bukau and Horwich, 1998; Burston and Walter, 2005; Thirumalai and Lorimer, 2001); the 90 kDa Hsps, e.g. Hsp90, HptG (Pearl and Prodromou, 2006; Richter et al., 2005), and finally the 100 kDa Hsps, e. g. yeast Hsp104, *E.coli* ClpB (Mogk and Bukau, 2004; Bosl et al., 2006). All five classes are thought to protect cells from the damaging effects of stress conditions by suppressing or resolving protein aggregation. The molecular mechanisms of these chaperone families however, are not resolved yet. Interestingly, some chaperone families recognize a broad spectrum of substrates whereas other seem to be rather specific making it hard to predict which chaperones are needed to fold a particular protein.

The chaperone studied best is the *E. coli* GroE complex, consisting of 14 GroEL subunits, arranged in a cylinder of two heptameric rings where the co-chaperone GroES, also a heptameric ring, binds to the poles. Non-native proteins chains in a molecular size range from 10 to 60 kDa can be bound in the central cavity of the GroEL cylinder polypeptide (Bukau et al., 2006; Mayhew et al., 1996; Fenton and Horwich, 2003; Burston and Walter, 2005). In the presence of ATP, the encapsulated substrate is released and dissociation of GroES is triggered by ATP hydrolysis.

In contrast to the ring-shaped structures of GroE, the Hsp70 chaperones exert their ATP-dependent chaperone acivities as monomers. Generally, the ubiquitous Hsp70 proteins are involved in a wide range of cellular processes, including *de novo* protein folding, e. g. interacting with the native chain already on the ribosome (Bukau et al., 2000) and degradation of unstable proteins (Dougan et al., 2002). The common function of Hsp70 in these processes appears to be the binding of short hydrophobic segments in partially folded polypeptides, thereby preventing aggregation and controlling the folding process. Similar to GroE, the most extensively studied representative of this chaperone group is the bacterial DnaK protein. DnaK, and also many other Hsp70 chaperones, interacts with two other classes of partner proteins (DnaJ and GrpE proteins) that are necessary for the regulation of its ATP-dependent chaperone cycle. ATP binding and hydrolysis allows the DnaK protein to shift between high and low affinity states for substrate molecules. The co-chaperone DnaJ (Hsp40) stimulates ATP hydrolysis, while the nucleotide exchange factor GrpE facilitates the exchange of ADP against ATP (Bukau and Horwich, 1998).

In the last years, the Hsp100/Clp family of chaperones came into the focus of chaperone research. These chaperones play important cellular roles in thermotolerance, regulation of genetic competence, transposition, proteolysis, and control of protein-based genetic elements (prion) in different organisms (Mogk and Bukau, 2004; Bosl et al., 2006). These seemingly unrelated processes are unified by a common remarkable biochemical mechanism: the proteins promote the disassembly of aggregated proteins and higher-order protein complexes. Several biochemical properties are shared by Hsp100 proteins and are required for their biological functions, suggesting that these properties are important for protein disassembly. These properties include an ATP-hydrolyzing activity and the ability to self-assemble into oligomers, primarily hexamers (Schirmer et al., 2001).

Another variation on the theme of protecting proteins from irreversible aggregation by reversible interaction with specialized folding factors is represented by the small heat-shock proteins (sHsps). These proteins have been found in almost all organisms studied so far. sHsps have been included in the class of molecular chaperones because they specifically bind to unfolded proteins *in vitro* and prevent their aggregation in an ATP-independent way (Jakob and Buchner, 1994; Van Montfort et al., 2002; Haslbeck et al., 2005a; Haslbeck, 2006). Compared to the molecular chaperones discussed above, sHsps have a remarkable binding capacity. They are able to bind a large number of non-native proteins. Furthermore, there seems to be no restriction on the size of proteins that can be bound (Haslbeck et al., 2004; Basha et al., 2004; Haslbeck et al., 2004; Basha et al., 2004). Bacterial sHsps were found to be tightly associated with inclusion bodies which are formed during high level protein expression. In *E. coli*, the two endogenous sHsps were therefore termed inclusion body protein A and B (IbpA and IbpB; Allen et al., 1992; Kuczynska-Wisnik et al., 2002). In comparison to the other classes of chaperones, they show a characteristic heterogeneity in sequence and size. Their common trait is a conserved ~ 90-residue C-terminal domain, the so-called "α-crystallin domain", which refers to the most prominent family member, the eye-lens protein α-crystallin (Horwitz, 2003; De Jong et al., 1998). This region is flanked by an N-terminal hydrophobic region, quite divergent in sequence and length and a short C-terminal extension. In addition to sequence homology, both sHsps and α-crystallins form large oligomeric complexes, although complex size can vary substantially in a size range of 12 to 42 subunits (Haslbeck et al., 2005a). Recent studies expanded the picture of the chaperone network centered around the sHsp-substrate protein complexes. Studies in *E. coli* and *S. cerevisiae* revealed a functional triade of chaperones which involves Hsp70 and members of the Hsp100 family (Matuszewska et al., 2005; Mogk et al., 2003a; Haslbeck et al., 2005b; Cashikar et al., 2005). It was shown that the Hsp70 and Hsp100 proteins are able to efficiently promote the refolding of substrates from both soluble sHsp/substrate complexes and from insoluble substrate aggregates.

The Hsp90 chaperone family is highly conserved between prokaryotes and eukaryotes, but eukaryotes in particular have evolved this family to a far more complex chaperone machinery than GroE and DnaK (Pearl and Prodromou, 2006; Richter et al., 2005). Nevertheless, Hsp90 is one of the most abundant proteins in the eukaryotic cell under normal conditions (Lindquist and Craig, 1988). In eukaryotes, Hsp90 seems to be involved in the conformational regulation of key proteins of signaling pathways such as steroid receptors and kinases under physiological conditions and in protein refolding. Stable interaction with folding intermediates is conferred by the large number of partner proteins with which eukaryotic Hsp90 is known to cooperate. Eukaryotic Hsp90 seems to perform at least part of its function in complex with Hsp70, large prolyl isomerases and p23, the latter two proteins being important partners of Hsp90 which have been characterized as chaperone-like proteins *in vitro* (Freeman et al., 2000; Freeman et al., 1996; Bose et al., 1996).

Besides these classical chaperone families especially in the cytosol of *E.coli* (which is still the source for the most widespread lysate types used in *in vitro* transcription / translation) two more interesting molecular chaperones exist. One is trigger factor, a chaperone directly associated with the ribosome and assisting the folding of up to 80% of the newly synthesized nascent chains after leaving the exit tunnel of the ribosome (Deuerling and Bukau, 2004). The second is Hsp33, a redox-regulated chaperone activated under oxidative stress conditions (Winter and Jakob, 2004).

## 5. Molecular Chaperones to Enhance *in vitro* Transcription / Translation

Regarding protein folding, *in vitro* translation systems highly resemble the situation in the cell. Most *in vitro* transcription / translation systems are based on bacterial or eukaryotic cell lysates and the protein concentration in the reaction is still relatively high. Thus, it is not to surprising that in many cases aggregation of the synthesized target protein has been observed during *in vitro* transcription / translation. Like in the cell, the folding of newly synthesized proteins seems to depend on the protective function of molecular chaperones. However, in contrast to a functional cell, the amount of molecular chaperones is not adjusted to the rate of protein synthesis automatically.

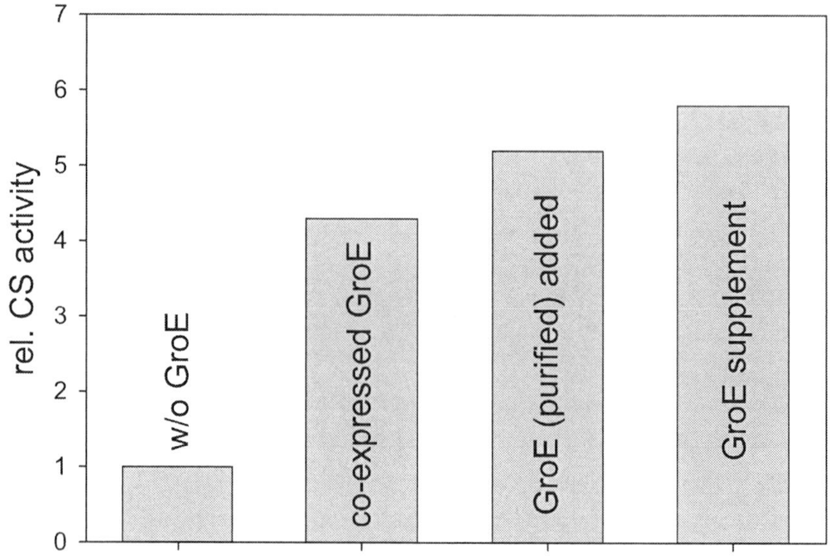


As an example for the increase of the amount of active target proteins after expression in the presence of supplemented chaperones, the expression of citrate synthase (CS) was compared using different forms of GroE supplementation. The amount of active CS after expression without (w/o) additional chaperone was normalized to 1. The co-expression, the addition of purified GroE and the use of GroE supplement (Roche Diagnostics, Mannheim, Germany) showed a comparable increase in the amount of active CS (Haslbeck et al., 2003; Haslbeck et al., 2005b).

Figure 5. The supplementation of GroE enhances the yield of correctly folded and active target protein

In the last years it has been demonstrated for several target proteins that the supplementation of GroE, the DnaK system or other chaperones to the *in vitro* translation reaction resulted in a significant increase of solubility of the target protein (Jiang et al., 2002; Stemp et al., 2005; Nakano et al., 2004). A higher concentration of chaperones in cell lysates can be achieved by the addition of purified chaperones. However, the use of purified chaperones is expensive. In most cases crude chaperone preparations are sufficient (GroE supplement; Roche Diagnostics, Mannheim, Germany). A second possibility to influence target protein folding is the parallel expression of chaperones and target protein in one reaction. In figure 5, the experimental approaches are compared for the expression of porcine mitochondrial citrate synthase (CS) as the target protein. As indicated, each approach to enhance the GroE concentration is able to increase the amount of active, protein significantly. Surprisingly, the yields obtained in the case of the co-expression of the GroEL/ES system resemble those after addition of purified GroEL and GroES.

In most cases, the supplemented chaperones increase the solubility of the target proteins. Interestingly, the supplementation of GroE and DnaK is also helpful when eukaryotic, secreted or disulfide-bridged proteins are synthesized (Jiang et al., 2002; Stemp et al., 2005). In the case of the already discussed expression of urokinase in a system optimized for disulfide formation, the addition of GroE helped efficiently to increase the solubility of urokinase (Kim et al., 2003).

However, to enhance the soluble and active expression of complex eukaryotic proteins like kinases, the supplementation of the reaction with specialized eukaryotic chaperones like Hsp90 might be necessary. Or stated the other way round, it seems possible to program the bacterial lysate for the special folding needs of eukaryotic proteins.

# 6. Studying Protein Folding in *in vitro* Translation Systems

Chaperones provide a useful toolkit to enhance protein synthesis in *in vitro* transcription / translation systems for biotechnological purposes such as screening and high throughput protein expression.

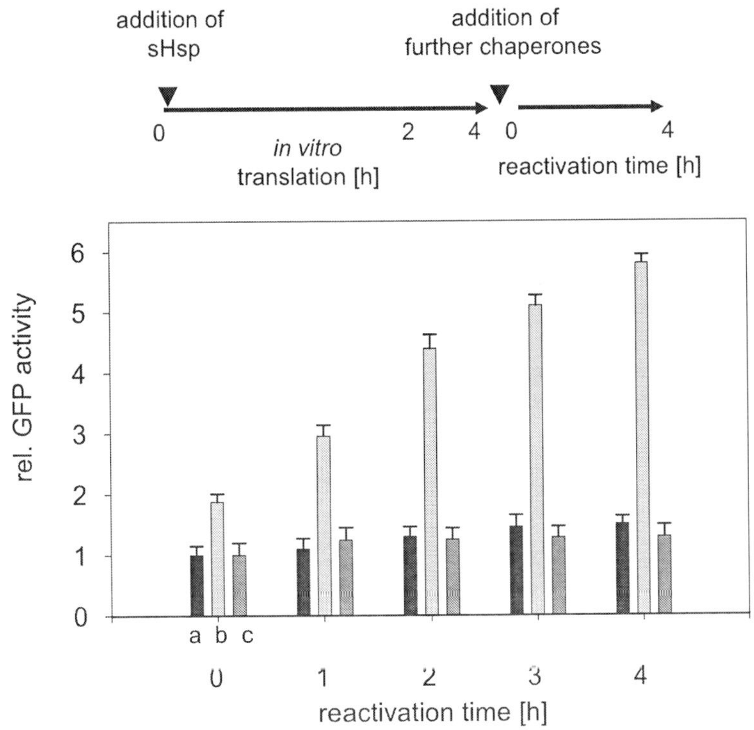


*In vitro* transcription / translation reactions were carried out using a $RTS_{500}$ devise at 30°C, 120 rpm stirring speed for 24 h. In each reaction, GFP was expressed from 10 μg of template DNA. As demonstrated in the experimental flow chart, samples were split after synthesis in the presence (a) of 300 nM sHsp (Hsp26) and a combination of 300 nM Hsp70 and Hsp100 chaperone systems (Ssa1 and Hsp104 respectively) was added (b). Reactivation of GFP was monitored determining the GFP fluorescence at 503 nm at the time points indicated. In (c) Hsp26 was absent during *in vitro* transcription / translation and 300 nM of the Hsp70/Hsp100 combination was added after the reaction (Haslbeck et al., 2005b).

Figure 6. Hsp70 and Hsp100 chaperones promote the folding of substrates captured by sHsps

On the other hand, several recent mechanistic studies on chaperone function made use of *in vitro* translation systems to gain insight into specific functional aspects. In a recent example *in vitro* transcription / translation and GFP as a model substrate was used to determine how substrates trapped by sHsps during heat stress are released (Haslbeck et al.,

2005b). The yeast sHsp Hsp26 suppresses the aggregation of GFP by forming stable complexes with it. To determine if the bound GFP is not trapped in dead-end complexes GFP was expressed in an *in vitro* transcription / translation system. Only 15-30% of GFP reached the native state after *in vitro* synthesis and the remainder forms insoluble aggregates lacking fluorescence. In the presence of Hsp26, complexes with the aggregating GFP are formed in the transcription / translation reaction. After completion of the translation reaction, the addition of the yeast Hsp70 and Hsp104 system lead to an efficient release of GFP (Fig. 6). In contrast, the addition of *E. coli* GroE did not result in a significant increase of GFP activity. When the *in vitro* transcription / translation reaction was performed in the absence of the sHsp, the subsequent addition of chaperones did not trigger any refolding of GFP. Together with other data, these studies increased the mechanistic understanding of how sHsps and other chaperones work together in a cascade-like chaperone network where substrates are handed over from the sHsps to Hsp100 and Hsp70 chaperones while recovery from cytosolic aggregates (Haslbeck et al., 2005b; Cashikar et al., 2005; Mogk et al., 2003b).

In an other study, the RTS was used to demonstrate that for efficient expression and folding of eukaryotic actin only the eukaryotic Hsp60 homologue CCT/TRiC is necessary and sufficient (Stemp et al., 2005). Recombinant expression of actin in bacteria or in the RTS results in non-native species tending to aggregate. Supplementing different chaperones in the RTS, Barral and colleagues were able to demonstrate that GroE and DnaK as well as CCT/TRiC were able to prevent the aggregation of actin, but only CCT/TRiC supported the correct folding of actin (Stemp et al., 2005). Thus, this combination of a bacterial *in vitro* translation with CCT/TRiC-assisted folding allowed the detailed analysis of the components necessary for efficient folding of eukaryotic actin.

But not only the influences of chaperones on the folding of target proteins can be studied, *in vitro* translation systems can also be utilized to study the folding of the target protein. When protein folding is analyzed in the test tube, intrinsic Trp fluorescence or labeling of the protein with fluorescence dyes is often used to monitor folding processes or interactions of protein domains. Of course it is also possible to add fluorescence-labeled amino acids in the *in vitro* translation systems. Even more impressive is the exchange of the canonical tryptophane by engineered variants. For example, Sengua et al. studied the homo- and heterodimer formation of *E. coli* RscR protein, a key regulator in enteric and plant pathogenesis (Ross et al., 1997). The experiments showed that the replacement of the natural Trp residue in RscB by various Trp analogues as well as high level production of the modified RscB was possible. The isolated RscB variants proved to be suitable to study protein interactions by fluorescence spectroscopy.

A very recent study proofed that *in vitro* translation is even possible on a chip-based technology allowing the synthesis of proteins directly from a microarray spotted with template DNA (Nakano et al., 2004; Jackson et al., 2004; Ramachandran et al., 2006). On microarrays, the enzymatic properties as well as protein-protein interactions can be studied for thousands of purified proteins in parallel. Thus, they provide a powerful platform for the high-throughput analysis of protein function. Traditionally the method requires production and purification of the proteins followed by printing them on the chip. A novel approach, called nucleic acid programmable protein array (NAPPA) allows the synthesis of the proteins directly on the chip (Ramachandran et al., 2006). NAPPA starts with the spotting of the template DNA on the chip followed immediately by *in vitro* transcription / translation. The expressed proteins are captured and oriented at the site of synthesis by their N- or the C-terminus. This method allows the just-in-time synthesis of the proteins for assaying. Thus, in comparison to conventional printing technologies, NAPPA minimizes most concerns about protein stability and integrity.

# 7. Conclusion

Coupled *in vitro* transcriptions / translation systems will be a key technology of the post-genome era. For the functional analysis of the whole set of human proteins, high throughput technologies allowing the fast and effective expression and purification of many proteins in parallel are required. Because *in vitro* transcription / translation is possible by directly utilizing PCR-products, and thus omitting cloning procedures, it provides an important advantage over conventional *in vivo* expression systems. Even more, the complete automation from PCR to purified protein seems possible. A further advantage is that the *in vitro* system is an open system allowing the specific modification of the reaction mixture. Thus, the expression of very complex, toxic or even disulfide-bonded proteins is possible. The targeted addition of chaperones or other folding catalysts provides the opportunity for fast and parallel optimization of protein synthesis. This system seems also very promising in the analysis of diseases caused by the incorrect folding of a protein. Here, combined studies of the influences of chaperones and potential therapeutic agents could provide new therapeutical concepts.

## Acknowledgement

We thank Roche Diagnostics and the Bayerische Forschungsstiftung for financial support.

# References

[1] Allen,S.P., Polazzi,J.O., Gierse,J.K., and Easton,A.M. (1992). Two novel heat shock genes encoding proteins produced in response to heterologous protein expression in Eschcrichia coli. *J. Bacteriol.* **174**, 6938-6947.

[2] Baranov,V.I., Morozov,I.Y., Ortlepp,S.A., and Spirin,A.S. (1989). Gene expression in a cell-free system on the preparative scale. *Gene* **84**, 463-466.

[3] Baranov,V.I. and Spirin,A.S. (1993). Gene expression in cell-free system on preparative scale. *Methods Enzymol.* **217**, 123-142.

[4] Basha,E., Lee,G.J., Breci,L.A., Hausrath,A.C., Buan,N.R., Giese,K.C., and Vierling,E. (2004). The identity of proteins associated with a small heat shock protein during heat stress in vivo indicates that these chaperones protect a wide range of cellular functions. *J. Biol. Chem.* **279**, 7566-7575.

[5] Betton,J.M. (2003). Using Maltose-Bnding Protein Fragment Complementation to Probe Protein-Protein Interactions by co expression in the RTS System. In: *Cell-Free Protein Expression,* Swartz,J.R., ed., Springer, Berlin, pp. 143-150.

[6] Betton,J.M. and Hofnung,M. (1994). In vivo assembly of active maltose binding protein from independently exported protein fragments. *EMBO J.* **13**, 1226-1234.

[7] Borosook,H., Deasy,C.L., Haagen-Smit,A.J., Keighley,G., and Lowy,P.H. (1950). The uptake in vitro of C14-labeled glycine, l-leucine, and l-lysine by different components of guinea pig liver homogenate. *J. Biol. Chem.* **184**, 529-543.

[8] Bose,S., Weikl,T., Bugl,H., and Buchner,J. (1996). Chaperone function of Hsp90-associated proteins. *Science* **274**, 1715-1717.

[9] Bosl,B., Grimminger,V., and Walter,S. (2006). The molecular chaperone Hsp104-A molecular machine for protein disaggregation. *J. Struct. Biol.* Epub ahead of print

[10] Bosserhoff,A.K. (2003). Recombinant Expression of Functional Active MIA Protein for Mutation analysis Using the RTS System. In: *Cell-Free Protein Expression*, Swartz,J.R., ed., Springer, Berlin, pp. 173-180.

[11] Bujard,H., Gentz,R., Lanzer,M., Stueber,D., Mueller,M., Ibrahimi,I., Haeuptle,M.T., and Dobberstein,B. (1987). A T5 promoter-based transcription-translation system for the analysis of proteins in vitro and in vivo. *Methods Enzymol.* **155**, 416-433.

[12] Bukau,B., Deuerling,E., Pfund,C., and Craig,E.A. (2000). Getting newly synthesized proteins into shape. *Cell* **101**, 119-122.

[13] Bukau,B. and Horwich,A.L. (1998). The Hsp70 and Hsp60 chaperone machines. *Cell* **92**, 351-366.

[14] Bukau,B., Weissman,J., and Horwich,A. (2006). Molecular chaperones and protein quality control. *Cell* **125**, 443-451.

[15] Burston,S.G. and Walter,S. (2005). The E. coli GroE Chaperone. In: *Protein Folding Handbook*, Buchner,J. and Kiefhaber,T., eds., Wiley VCH, Weinheim, pp. 699-724.

[16] Cashikar,A.G., Duennwald,M.L., and Lindquist,S.L. (2005). A chaperone pathway in protein disaggregation: Hsp26 alters the nature of protein aggregates to facilitate reactivation by Hsp104. *J. Biol. Chem.* **280**, 23869-23875.

[17] Chao,F.C. and Schachman,H.K. (1956). The isolation and characterization of a macro-molecular ribonucleoprotein from yeast. *Arch. Biochem. Biophys.* **61**, 220-230.

[18] Christendat,D., Yee,A., Dharamsi,A., Kluger,Y., Gerstein,M., Arrowsmith,C.H., and Edwards,A.M. (2000a). Structural proteomics: prospects for high throughput sample preparation. *Prog. Biophys. Mol. Biol.* **73**, 339-345.

[19] Christendat,D., Yee,A., Dharamsi,A., Kluger,Y., Savchenko,A., Cort,J.R., Booth,V., Mackereth,C.D., Saridakis,V., Ekiel,I., Kozlov,G., Maxwell,K.L., Wu,N., McIntosh,L.P., Gehring,K., Kennedy,M.A., Davidson,A.R., Pai,E.F., Gerstein,M., Edwards,A.M., and Arrowsmith,C.H. (2000b). Structural proteomics of an archaeon. *Nat. Struct. Biol.* **7**, 903-909.

[20] Craig,D., Howell,M.T., Gibbs,C.L., Hunt,T., and Jackson,R.J. (1992). Plasmid cDNA-directed protein synthesis in a coupled eukaryotic in vitro transcription-translation system. *Nucleic Acids Res.* **20**, 4987-4995.

[21] D'Aniello,A., D'Onofrio,G., Pischetola,M., D'Aniello,G., Vetere,A., Petrucelli,L., and Fisher,G.H. (1993). Biological role of D-amino acid oxidase and D-aspartate oxidase. Effects of D-amino acids. *J. Biol. Chem.* **268**, 26941-26949.

[22] De Jong,W.W., Caspers,G.J., and Leunissen,J.A. (1998). Genealogy of the alpha-crystallin--small heat-shock protein superfamily. *Int. J. Biol. Macromol.* **22**, 151-162.

[23] Deuerling,E. and Bukau,B. (2004). Chaperone-assisted folding of newly synthesized proteins in the cytosol. Crit Rev. Biochem. *Mol. Biol.* **39**, 261-277.

[24] DeVries,J.K. and Zubay,G. (1967). DNA-directed peptide synthesis. II. The synthesis of the alpha-fragment of the enzyme beta-galactosidase. *Proc. Natl. Acad. Sci. U. S. A* **57**, 1010-1012.

[25] Dieckman,L., Gu,M., Stols,L., Donnelly,M.I., and Collart,F.R. (2002). High throughput methods for gene cloning and expression. *Protein Expr. Purif.* **25**, 1-7.

[26] Dieckman,L.J., Hanly,W.C., and Collart,E.R. (2006). Strategies for high-throughput gene cloning and expression. *Genet. Eng (N. Y. )* **27**, 179-190.

[27] Dill,K.A. and Chan,H.S. (1997). From Levinthal to pathways to funnels. *Nat. Struct. Biol* **4**, 10-19.

[28] Dougan,D.A., Mogk,A., and Bukau,B. (2002). Protein folding and degradation in bacteria: to degrade or not to degrade? That is the question. *Cell Mol. Life Sci.* **59**, 1607-1616.

[29] Fenton,W.A. and Horwich,A.L. (2003). Chaperonin-mediated protein folding: fate of substrate polypeptide. *Q. Rev. Biophys.* **36**, 229-256.

[30] Freeman,B.C., Felts,S.J., Toft,D.O., and Yamamoto,K.R. (2000). The p23 molecular chaperones act at a late step in intracellular receptor action to differentially affect ligand efficacies. *Genes Dev.* **14**, 422-434.

[31] Freeman,B.C., Toft,D.O., and Morimoto,R.I. (1996). Molecular chaperone machines: chaperone activities of the cyclophilin Cyp-40 and the steroid aporeceptor-associated protein p23. *Science* **274**, 1718-1720.

[32] Green,R.H., Glick,B.R., and Ganoza,M.C. (1985). Requirements for in vitro reconstruction of protein synthesis. *Biochem. Biophys. Res. Commun.* **126**, 792-798.

[33] Guba,M., Bosserhoff,A.K., Steinbauer,M., Abels,C., Anthuber,M., Buettner,R., and Jauch,K.W. (2000). Overexpression of melanoma inhibitory activity (MIA) enhances extravasation and metastasis of A-mel 3 melanoma cells in vivo. *Br. J. Cancer* **83**, 1216-1222.

[34] Guignard,L., Ozawa,K., Pursglove,S.E., Otting,G., and Dixon,N.E. (2002). NMR analysis of in vitro-synthesized proteins without purification: a high-throughput approach. *FEBS Lett.* **524**, 159-162.

[35] Hartl,F.U. and Hayer-Hartl,M. (2002). Molecular chaperones in the cytosol: from nascent chain to folded protein. *Science* **295**, 1852-1858.

[36] Haslbeck,M. (2006). Recombinant expression and in vitro refolding of the yeast small heat shock protein Hsp42. *Int. J. Biol. Macromol.* **38**, 107-114.

[37] Haslbeck,M., Braun,N., Stromer,T., Richter,B., Model,N., Weinkauf,S., and Buchner,J. (2004). Hsp42 is the general small heat shock protein in the cytosol of Saccharomyces cerevisiae. *EMBO J.* **23**, 638-649.

[38] Haslbeck,M., Franzmann,T., Weinfurtner,D., and Buchner,J. (2005a). Some like it hot: the structure and function of small heat-shock proteins. *Nat. Struct. Mol. Biol.* **12**, 842-846.

[39] Haslbeck,M., Miess,A., Stromer,T., Walter,S., and Buchner,J. (2005b). Disassembling protein aggregates in the yeast cytosol: The cooperation of Hsp26 with Ssa1 and Hsp104. *J. Biol. Chem.* **280**, 23861-23868.

[40] Haslbeck,M., Schuster,I., and Grallert,H. (2003). GroE-dependent expression and purification of pig heart mitochondrial citrate synthase in Escherichia coli. *J. Chromatogr. B Analyt. Technol. Biomed. Life Sci.* **786**, 127-136.

[41] Hoffmann,M., Nemetz,C., Madin,K., and Buchberger,B. (2004). Rapid translation system: a novel cell-free way from gene to protein. *Biotechnol. Annu. Rev.* **10**, 1-30.

[42] Hofweber,R., Horn,G., Langmann,T., Balbach,J., Kremer,W., Schmitz,G., and Kalbitzer,H.R. (2005). The influence of cold shock proteins on transcription and translation studied in cell-free model systems. *FEBS J.* **272**, 4691-4702.

[43] Horwich,A.L., Farr,G.W., and Fenton,W.A. (2006). GroEL-GroES-mediated protein folding. Chem. Rev. **106**, 1917-1930.

[44] Horwitz,J. (2003). Alpha-crystallin. *Exp. Eye Res.* **76**, 145-153.

[45] Hurley,J.H., Anderson,D.E., Beach,B., Canagarajah,B., Ho,Y.S., Jones,E., Miller,G., Misra,S., Pearson,M., Saidi,L., Suer,S., Trievel,R., and Tsujishita,Y. (2002). Structural genomics and signaling domains. *Trends Biochem. Sci.* **27**, 48-53.

[46] Hyun,B.J., Rubini,M., Jung,G., Wiegand,G., Seifert,M.H., Azim,M.K., Kim,J.S., Zumbusch,A., Holak,T.A., Moroder,L., Huber,R., and Budisa,N. (2003). Expansion of the genetic code enables design of a novel "gold" class of green fluorescent proteins. *J. Mol. Biol.* **328**, 1071-1081.

[47] Jackson,A.M., Boutell,J., Cooley,N., and He,M. (2004). Cell-free protein synthesis for proteomics. *Brief. Funct. Genomic. Proteomic.* **2**, 308-319.

[48] Jaenicke,R. (1998). Protein self-organization in vitro and in vivo: partitioning between physical biochemistry and cell biology. *Biol. Chem.* **379**, 237-243.

[49] Jaenicke,R. and Seckler,R. (1997). Protein misassembly *in vitro*. *Adv. Protein Chem.* **50**, 1-59.

[50] Jakob,U. and Buchner,J. (1994). Assisting spontaneity: the role of Hsp90 and small Hsps as molecular chaperones. *Trends Biochem. Sci.* **19**, 205-211.

[51] Jiang,X., Ookubo,Y., Fujii,I., Nakano,H., and Yamane,T. (2002). Expression of Fab fragment of catalytic antibody 6D9 in an Escherichia coli in vitro coupled transcription/translation system. *FEBS Lett.* **514**, 290-294.

[52] Johansson,L., Gafvelin,G., and Arner,E.S. (2005). Selenocysteine in proteins-properties and biotechnological use. *Biochim. Biophys. Acta* **1726**, 1-13.

[53] Keller,E.B. and Zamecnik,P.C. (1956). The effect of guanosine diphosphate and triphosphate on the incorporation of labeled amino acids into proteins. *J. Biol. Chem.* **221**, 45-59.

[54] Kigawa,T., Muto,Y., and Yokoyama,S. (1995). Cell-free synthesis and amino acid-selective stable isotope labeling of proteins for NMR analysis. *J. Biomol. NMR* **6**, 129-134.

[55] Kigawa,T., Yabuki,T., Matsuda,N., Matsuda,T., Nakajima,R., Tanaka,A., and Yokoyama,S. (2004). Preparation of Escherichia coli cell extract for highly productive cell-free protein expression. *J. Struct. Funct. Genomics* **5**, 63-68.

[56] Kigawa,T., Yabuki,T., Yoshida,Y., Tsutsui,M., Ito,Y., Shibata,T., and Yokoyama,S. (1999). Cell-free production and stable-isotope labeling of milligram quantities of proteins. *FEBS Lett.* **442**, 15-19.

[57] Kigawa,T., Yamaguchi-Nunokawa,E., Kodama,K., Matsuda,T., Yabuki,T., Matsuda,N., Ishitani,R., Nureki,O., and Yokoyama,S. (2002). Selenomethionine incorporation into a protein by cell-free synthesis. *J. Struct. Funct. Genomics* **2**, 29-35.

[58] Kim,D.M. and Swartz,J.R. (1999). Prolonging cell-free protein synthesis with a novel ATP regeneration system. *Biotechnol. Bioeng.* **66**, 180-188.

[59] Kim,K.M., Fernholz E., and Swartz,J.R. (2003). Cell-Free Expression of Proteins Containing Multiple Disulfide Bonds. In: *Cell-Fre Protein Expression*, Swartz,J.R., ed., Springer, Berlin, pp. 125-132.

[60] Kim,S.H. (2000). Structural genomics of microbes: an objective. *Curr. Opin. Struct. Biol.* **10**, 380-383.

[61] Kobayashi,T., Sakamoto,K., Takimura,T., Sekine,R., Kelly,V.P., Kamata,K., Nishimura,S., and Yokoyama,S. (2005). Structural basis of nonnatural amino acid recognition by an engineered aminoacyl-tRNA synthetase for genetic code expansion. *Proc. Natl. Acad. Sci. U. S. A* **102**, 1366-1371.

[62] Kuczynska-Wisnik,D., Kedzierska,S., Matuszewska,E., Lund,P., Taylor,A., Lipinska,B., and Laskowska,E. (2002). The Escherichia coli small heat-shock proteins IbpA and IbpB prevent the aggregation of endogenous proteins denatured in vivo during extreme heat shock. *Microbiology* **148**, 1757-1765.

[63] Kung,H.F., Redfield,B., Treadwell,B.V., Eskin,B., Spears,C., and Weissbach,H. (1977). DNA-directed in vitro synthesis of beta-galactosidase. Studies with purified factors. *J. Biol. Chem.* **252**, 6889-6894.

[64] Leconte,A.M., Chen,L., and Romesberg,F.E. (2005). Polymerase evolution: efforts toward expansion of the genetic code. *J. Am. Chem. Soc.* **127**, 12470-12471.

[65] Lehmann,D. and Thiesen,H.J. (2003). Optimization of Cell-Free Expression of FAD-Dependent D-Amino Acid Oxidase. In: *Cell-Free Protein Expression*, Swartz,J.R., ed., Springer, Berlin, pp. 151-158.

[66] Lindquist,S. and Craig,E.A. (1988). The heat-shock proteins. *Annu. Rev Genet.* **22**, 631-677.

[67] Littlefield,J.W., Keller,E.B., Gross,J., and Zamecnik,P.C. (1955). Studies on cytoplasmic ribonucleoprotein particles from the liver of the rat. *J. Biol. Chem.* **217**, 111-123.

[68] Lorenz,P. and Thiesen,H.J. (2003). In vitro Translation of KRAB Zinc Finger Transcriptional Repressor Proteins and Their Interaction with Their TIF1ß Co-Repressor. In: *Cell Free Protein Expression,* Swartz,J.R., ed., Springer, Berlin, pp. 151-158.

[69] Madin,K., Sawasaki,T., Ogasawara,T., and Endo,Y. (2000). A highly efficient and robust cell-free protein synthesis system prepared from wheat embryos: plants apparently contain a suicide system directed at ribosomes. *Proc. Natl. Acad. Sci. U. S. A* **97**, 559-564.

[70] Martin,G.A., Kawaguchi,R., Lam,Y., DeGiovanni,A., Fukushima,M., and Mutter,W. (2001). High-yield, in vitro protein expression using a continuous-exchange, coupled transcription/ translation system. *Biotechniques* **31**, 948-3.

[71] Matthaei,J.H. and Nirenberg,M.W. (1961). Characteristics and stabilization of DNAase-sensitive protein synthesis in E. coli extracts. *Proc. Natl. Acad. Sci. U. S. A* **47**, 1580-1588.

[72] Matuszewska,M., Kuczynska-Wisnik,D., Laskowska,E., and Liberek,K. (2005). The small heat shock protein IbpA of Escherichia coli cooperates with IbpB in stabilization of thermally aggregated proteins in a disaggregation competent state. *J. Biol. Chem.* **280**, 12292-12298.

[73] Mayhew,M., da Silva,A.C., Martin,J., Erdjument-Bromage,H., Tempst,P., and Hartl,F.U. (1996). Protein folding in the central cavity of the GroEL-GroES chaperonin complex. *Nature* **379**, 420-426.

[74] McIntosh,L.P. and Dahlquist,F.W. (1990). Biosynthetic incorporation of 15N and 13C for assignment and interpretation of nuclear magnetic resonance spectra of proteins. *Q. Rev. Biophys.* **23**, 1-38.

[75] Mogk,A. and Bukau,B. (2004). Molecular chaperones: structure of a protein disaggregase. *Curr. Biol.* **14**, R78-R80.

[76] Mogk,A., Deuerling,E., Vorderwulbecke,S., Vierling,E., and Bukau,B. (2003a). Small heat shock proteins, ClpB and the DnaK system form a functional triade in reversing protein aggregation. *Molecular Microbiology* **50**, 585-595.

[77] Mogk,A., Schlieker,C., Friedrich,K.L., Schofeld,H.J., Vierling,E., and Bukau,B. (2003b). Refolding of substrates bound to small Hsps relies on a disaggregation reaction mediated most efficiently by ClpB/DnaK. *J. Biol. Chem.* **278**, 31033-31042.

[78] Nakano,H., Kawarasaki,Y., and Yamane,T. (2004). Cell-free protein synthesis systems: increasing their performance and applications. *Adv. Biochem. Eng Biotechnol.* **90**, 135-149.

[79] Nirenberg,M. (2004). Historical review: Deciphering the genetic code--a personal account. *Trends Biochem. Sci.* **29**, 46-54.

[80] Nirenberg,M.W. and Matthaei,J.H. (1961). The dependence of cell-free protein synthesis in E. coli upon naturally occurring or synthetic polyribonucleotides. *Proc. Natl. Acad. Sci. U. S. A* **47**, 1588-1602.

[81] Ozawa,K., Dixon,N.E., and Otting,G. (2005). Cell-free synthesis of 15N-labeled proteins for NMR studies. IUBMB. *Life* **57**, 615-622.

[82] Paulus,M., Haslbeck,M., and Watzele,M. (2004). RNA stem-loop enhanced expression of previously non-expressible genes. *Nucleic Acids Res.* **32**, e78.

[83] Pearl,L.H. and Prodromou,C. (2006). Structure and mechanism of the hsp90 molecular chaperone machinery. *Annu. Rev. Biochem.* **75**, 271-294.

[84] Ramachandran,N., Hainsworth,E., Demirkan,G., and LaBaer,J. (2006). On-chip protein synthesis for making microarrays. *Methods Mol. Biol.* **328**, 1-14.

[85] Richter,K., Meinlschmidt,B., and Buchner J. (2005). Hsp90: From Dispensable Heat Shock Protein to Global Player. In: *Protein Folding Handbook*, Buchner,J. and Kiefhaber,T., eds., Wiley VCH, Weinheim, pp. 768-857.

[86] Roberts,B.E., Gorecki,M., Mulligan,R.C., Danna,K.J., Rozenblatt,S., and Rich,A. (1975). Simian virus 40 DNA directs synthesis of authentic viral polypeptides in a linked transcription-translation cell-free system. *Proc. Natl. Acad. Sci. U. S. A* **72**, 1922-1926.

[87] Roberts,B.E. and Paterson,B.M. (1973). Efficient translation of tobacco mosaic virus RNA and rabbit globin 9S RNA in a cell-free system from commercial wheat germ. *Proc. Natl. Acad. Sci. U. S. A* **70**, 2330-2334.

[88] Ross,J.B., Szabo,A.G., and Hogue,C.W. (1997). Enhancement of protein spectra with tryptophan analogs: fluorescence spectroscopy of protein-protein and protein-nucleic acid interactions. *Methods Enzymol.* **278**, 151-190.

[89] Schirmer,E.C., Ware,D.M., Queitsch,C., Kowal,A.S., and Lindquist,S.L. (2001). Subunit interactions influence the biochemical and biological properties of Hsp104. *Proc. Natl. Acad. Sci. U. S. A* **98**, 914-919.

[90] Scott,A., Pantoja-Uceda,D., Koshiba,S., Inoue,M., Kigawa,T., Terada,T., Shirouzu,M., Tanaka,A., Sugano,S., Yokoyama,S., and Guntert,P. (2004). NMR assignment of the SH2 domain from the human feline sarcoma oncogene *FES. J. Biomol. NMR* **30**, 463-464.

[91] Shimizu,Y., Inoue,A., Tomari,Y., Suzuki,T., Yokogawa,T., Nishikawa,K. and Ueda,T. (2001) Cell-free translation reconstituted with purified components. *Nat Biotechnol.* **19**, 751-755.

[92] Siekevitz,P. and Zamecnik,P.C. (1981). Ribosomes and protein synthesis. *J. Cell Biol.* **91**, 53s-65s.

[93] Spirin,A.L. (2002a). Cell-Free Protein Synthesis. In: *Cell-Free Translation Systems*, Spirin,A.L., ed., Springer, Berlin, pp. 3-20.

[94] Spirin,A.L. (2002b). Continuous-Flow and Continuous-Exchange Cell-Free Translation systems and Reactors. In: *Cell-Free Translation Systems*, Spirin,A.L., ed., Springer, Berlin, pp. 91-107.

[95] Spirin,A.S. (2004). High-throughput cell-free systems for synthesis of functionally active proteins. *Trends Biotechnol.* **22**, 538-545.

[96] Spirin,A.S., Baranov,V.I., Ryabova,L.A., Ovodov,S.Y., and Alakhov,Y.B. (1988). A continuous cell-free translation system capable of producing polypeptides in high yield. *Science* **242**, 1162-1164.

[97] Staunton,D., Schlinkert,R., Zanetti,G., Colebrook,S.A., and Campbell,I.D. (2006). Cell-free expression and selective isotope labelling in protein NMR. *Magn Reson. Chem.* **44**, 2-9.

[98] Stemp,M.J., Guha,S., Hartl,F.U., and Barral,J.M. (2005). Efficient production of native actin upon translation in a bacterial lysate supplemented with the eukaryotic chaperonin TRiC. *Biol. Chem.* **386**, 753-757.

[99] Stueber,D., Ibrahimi,I., Cutler,D., Dobberstein,B., and Bujard,H. (1984). A novel in vitro transcription-translation system: accurate and efficient synthesis of single proteins from cloned DNA sequences. *EMBO J.* **3**, 3143-3148.

[100] Swartz,J.R., Jewett,M.C., and Woodrow,K.A. (2004). Cell free protein synthesis with prokaryotic combined transcription-translation. *Methods Mol. Biol.* **267**, 169-182.

[101] Thirumalai,D. and Lorimer,G.H. (2001). Chaperonin-mediated protein folding. *Annu Rev. Biophys. Biomol. Struct.* **30**, 245-269.

[102] Tissieres,A. and Watson,J.D. (1958). Ribonucleoprotein particles from Escherichia coli. *Nature* **182**, 778-780.

[103] Tissieres,A. and Watson,J.D. (1962). Breakdown of messenger RNA during in vitro amino acid incorporation into proteins. *Proc. Natl. Acad. Sci. U. S. A* **48**, 1061-1069.

[104] Van Montfort,R., Slingsby,C., and Vierling,E. (2002). Structure and function of the small heat shock protein/alpha-crystallin family of molecular chaperones. *Adv. Protein Chem.* **59**, 105-156.

[105] Vitkup,D., Melamud,E., Moult,J., and Sander,C. (2001). Completeness in structural genomics. *Nat. Struct. Biol.* **8**, 559-566.

[106] Voges,D., Watzele,M., Nemetz,C., Wizemann,S., and Buchberger,B. (2004). Analyzing and enhancing mRNA translational efficiency in an Escherichia coli in vitro expression system. *Biochem. Biophys. Res. Commun.* **318**, 601-614.

[107] Walter,S. and Buchner,J. (2002). Molecular chaperones-cellular machines for protein folding. *Angew. Chem. Int. Ed Engl.* **41**, 1098-1113.

[108] Winter,J. and Jakob,U. (2004). Beyond transcription--new mechanisms for the regulation of molecular chaperones. *Crit Rev. Biochem. Mol. Biol.* **39**, 297-317.

[109] Yee,A., Pardee,K., Christendat,D., Savchenko,A., Edwards,A.M., and Arrowsmith,C.H. (2003). Structural proteomics: toward high-throughput structural biology as a tool in functional genomics. *Acc. Chem. Res.* **36**, 183-189.

[110] Yin,G. and Swartz,J.R. (2004). Enhancing multiple disulfide bonded protein folding in a cell-free system. *Biotechnol. Bioeng.* **86**, 188-195.

[111] Yokoyama,S. (2003). Protein expression systems for structural genomics and proteomics. *Curr. Opin. Chem. Biol.* **7**, 39-43.

[112] Zarucki-Schulz,T., Jerez,C., Goldberg,G., Kung,H.F., Huang,K.H., Brot,N., and Weissbach,H. (1979). DNA-directed in vitro synthesis of proteins involved in bacterial transcription and translation. *Proc. Natl. Acad. Sci. U. S. A* **76**, 6115-6119.

[113] Zhao,X., Georgakaki,I.P., Miller,M.L., Mejia-Rodriguez,R., Chiang,C.Y., and Darensbourg,M.Y. (2002). Catalysis of H(2)/D(2) scrambling and other H/D exchange processes by [Fe]-hydrogenase model complexes. *Inorg. Chem.* **41**, 3917-3928.

[114] Zubay,G. (1973). In vitro synthesis of protein in microbial systems. *Annu. Rev. Genet.* **7**, 267-287.

In: Protein Folding-Misfolding... ISBN 1-60021-417-7
Editors: J.P. Zbilut and T. Scheibel, pp. 117-140 

*Chapter 5*

# PROTEIN STRUCTURE PREDICTION AND MOLECULAR FORCES

***Manfred J. Sippl***
Center of Applied Molecular Engineering, University of Salzburg,
Jakob Haringer Str.1, A-5020 Salzburg, Austria;
sippl@came.sbg.ac.at

**Abstract**

Our ability to predict protein structures from amino acid sequences depends on our understanding of molecular forces. The same applies to the validation of protein structures determined in the laboratory. The protein structures available in the public domain contain a number of deficiencies and inconsistencies. As long as we are unable to recognize and correct such errors on a regular basis we cannot completely trust the experimental results and we cannot have any confidence in predicted structures. Here we investigate the theory of intermolecular forces from the perspective of protein structure theory, comment on the fundamental ideas involved, discuss the difficulties encountered, and we provide examples to illustrate the current state of affairs in protein structure validation and prediction.

## 1. Introduction

For more than half a century the protein folding problem challenges our understanding of physical systems on the molecular level. There seems to be a general consent regarding the view that the protein folding problem is at least qualitatively understood and that we do not need to invent new physical concepts or theories to explain the phenomenon. On the other hand we are still unable to predict structures with any confidence and we are still far from the precision obtained in experimental structure determination.

In an attempt to assess the current state of affairs in protein structure prediction an unbiased observer is bound to get confused. On the one hand he finds that the literature of the last fifty years abounds with claims that the protein folding problem has been solved. And on the other hand he finds that recently published protein structures determined with high precision are flawed with an astonishing amount of errors and deficiencies. Our observer might wonder how this fits together and what he can trust and what not.

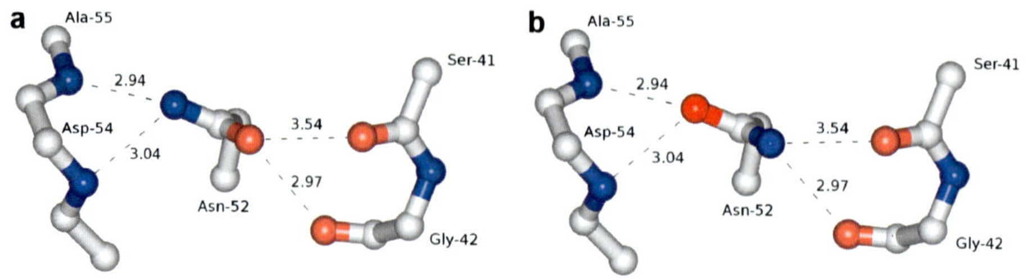


Figure 1. Asparagine Asn-52 of dethiobiotin synthetase (PDB code 1dad; resolution 1.6 Å). Here and in figures 2 and 3 dashed lines and the associated numbers represent distances between atoms (in Å). Atoms are colored by atom type: carbon, grey; oxygen, red; nitrogen, blue. The figures are generated using the program PyMOL (http://pymol.sourceforge.net). **a,** high energy rotamer as found in the crystal structure and **b,** correct low energy rotamer.

To be specific consider Figure 1. Here we see a small portion of the high resolution crystal structure of dethiobiotin synthetase determined to 1.6 Å resolution. The amide nitrogen of asparagine 52, which is a hydrogen bond donor, is in hydrogen bond contact with two other nitrogen atoms, whereas the oxygen atom of the same asparagine, a hydrogen bond acceptor, is in hydrogen bond contact with two other hydrogen bond acceptors. But this cannot be true. The hydrogen bond donors carry positive partial charges, whereas the hydrogen bond acceptors carry negative partial charges and hence the configuration found in the crystal structure is strongly repulsive violating elementary physico-chemical standards.

Obviously, we can correct the problem by flipping the amide plane as shown in Figure 1 (b). Now there are four ideal hydrogen bonds between the amide group of asparagine 52 and the surrounding hydrogen bond donors and acceptors. The difference in energy between the correct and the incorrect conformation of figure 1 is enormous. In the former we have four strong hydrogen bonds, in the latter these are substituted by highly unfavorable interactions creating the antidote of four hydrogen bonds. In fact, the energy difference between incorrect and correct configuration is so high that the protein disintegrates when this energy is released at once.

The problem seems easy enough to suspect that in experimentally determined protein structures this kind of error is the exception rather than the rule. But as a matter of fact one out of five amide groups of asparagine and glutamine side chains is found in the incorrect conformation. As a consequence the 30,000 protein structures currently available in the public domain contain more than half a million erroneous rotamers of asparagine and glutamine side chain amides.

The problem originates from the fact that electron densities around the amide nitrogen and oxygen atoms of asparagine and glutamine residues are quite similar so that the locations of these atoms can be determined to high precision but not their identity. Hence, the problem seems to be specific for X-ray analysis of protein crystals. However, the error rate in structures determined by nuclear magnetic resonance (NMR) is even slightly higher. The experimental data obtained from X-ray and NMR experiments are generally refined by various refinement protocols before the final coordinates are deposited and we have to conclude that the refinement protocols used are unable to correct unfavorable configurations or perhaps that they even introduce such errors.

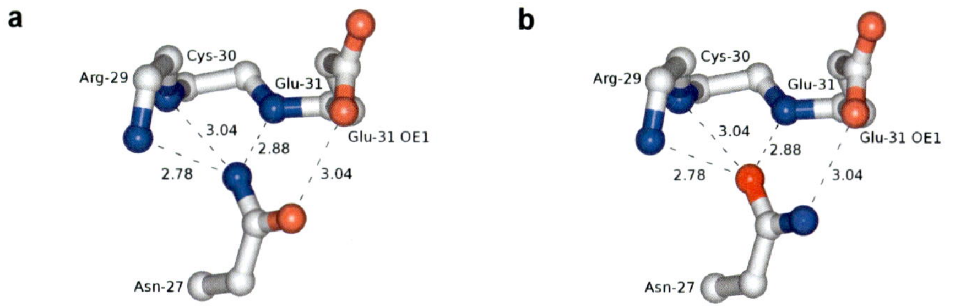


Figure 2. Asparagine Asn-27 of zinc metalloprotease (PDB code 1ezm; resolution 1.5 Å) and the molecular environment of this residues. **a,** high energy rotamer as found in the crystal structure and **b,** correct low energy rotamer.

Inspecting the example shown in figure 1 and similar cases (e.g. figures 2 and 3) and applying basic physico-chemical principles, we have not much difficulty in spotting the errors by eye, although there are other cases where this is not so obvious (e.g. see figure 1 of [49]). However, in presenting the case we have colored and oriented the atoms and labeled the atomic distances in ways to make the problem visible. In general such groups are hidden among thousands of atoms and they are not easy to find by visual inspection.

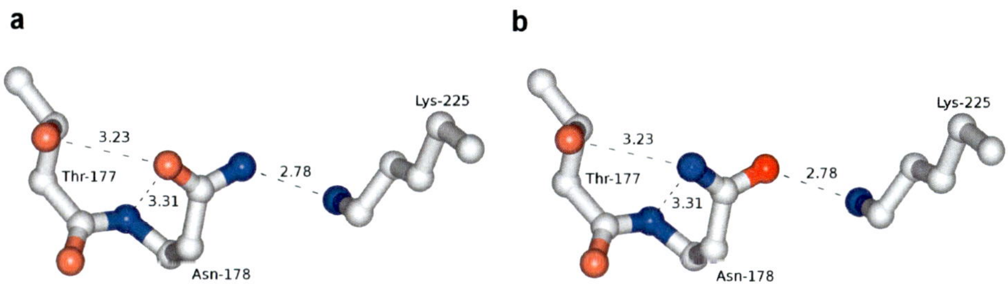


Figure 3. Asparagine Asn-178 of carbonic anhydrase (PDB code 2cba; resolution 1.54 Å). **a,** high energy rotamer as found in the crystal structure and **b,** correct low energy rotamer.

On the other hand it is generally assumed that the laws governing molecular interactions are known with sufficient quantitative precision so that such errors should be detectable by straightforward energy calculations. But again, this is more difficult than it sounds. First of all, there are many signs that tell us that the current models of molecular interactions are inadequate as we will discuss below. And second, X-ray analysis of protein crystal structures generally do not yield the positions of hydrogen atoms, but these atoms are required by most of the molecular force fields and energy functions currently in use. In particular, the energy of hydrogen bonds calculated by such molecular force fields strongly depends on the position of the proton which is shared between hydrogen bond donor and acceptor. Before any energy calculations can be done, the positions of the protons must be specified. The trouble is that the locations of the protons are not available from experiment and hence there is considerable freedom in the choice of proton positions. This choice may be guided by chemical intuition but some degree of arbitrariness necessarily remains. Hence, we are confronted with two difficult problems: the uncertainty in proton positions and the uncertainty in the quality of energy functions.

This introductory example shows that even our best structures contain an astonishing amount of errors. But this is only the tip of an iceberg. On a local scale there are similar problems with histidine rotamers, poorly resolved loops, disordered regions, etc. and on a global scale the public domain contains partially erroneous as well as completely misfolded structures. It is one of the major challenges of protein structure theory and the duty of curators of data bases to provide methods which can detect and correct such errors. Without convincing achievements in this direction attempts to predict protein structures remain elusive, we cannot have any confidence in predicted structures, and we even have to doubt the validity of experimental results.

Over the years there have been many successful predictions that seem to contradict such arguments and in particular the repeated CASP experiments have produced quite amazing predictions. But essentially all predictions reported in the literature rely heavily on sequence and structure data bases and expert knowledge and frequently a simple key word or sequence search yields an excellent prediction in the form of a related structure. Such results may turn out to be extremely useful in addressing biological questions, solving a related structure, studying biochemical mechanisms, and so on, but they leave the most important question unanswered, that is, is the structure correct?

The data mining aspects of protein structure prediction and related approaches are well covered by the CASP experiments and extensive reviews are available in the recent literature . There is no need to repeat this here. Instead we take the opportunity to reflect on some fundamental problems in protein structure theory. Our main focus are the molecular interactions found in proteins and in particular the hydrogen bond, ubiquitous in biological macromolecules. A commented summary of the bibliography is found in the last section of this report.

## 2. Molecular Forces in Quantum Mechanics

The problem of molecular forces and the cohesion of matter has a long and interesting history [33]. Today the definitive language of molecules is quantum mechanics and hence any approach to molecular interactions necessarily has to start with Schrödinger's equation and the associated wave functions. We will not go into details here, but we need a brief review of what quantum mechanical calculations are able to do today. This also sheds some light on the origin of the force fields currently used in structure prediction, molecular modeling, and molecular dynamics simulations of proteins and in the refinement protocols employed in protein structure determination by X-ray and NMR methods.

The problem with quantum mechanics generally is that either we have studied the subject in some detail so that we are fluent in the language, know how to construct and solve the Schrödinger equation, and have the necessary experience to interpret the results. Assimilation of the postulates and mastering the subject in some detail certainly requires a couple of years. If we are interested in proteins we usually cannot afford the time necessary to go into such detail. Hence, basic expressions of quantum mechanics, like the Schrödinger equation, are not very informative since an expert knows them by heart whereas they are unintelligible to the novice. But they are a necessary starting point as can be judged by any textbook on physics and chemistry written after 1926.

The basic recipe of molecular quantum mechanics goes like this: Write down the Schrödinger equation for the molecular system of interest. The resulting equation is always 'exact' but since we cannot solve the equation directly we have to apply an arsenal of elegant and ingenious techniques to find approximate solutions. The calculations are considered to be successful if the results resemble experimental data. If the discrepancy between theory and experiment is too large, we can go back and try other tricks or simplify the system. There is however, a certain danger to get lost in this loop.

To be specific, the Schrödinger equation is written in the compact notation

$$H\Psi = E\Psi, \tag{1}$$

where $H$ is the Hamiltonian, also called the energy operator, of the molecular system, $\Psi$ are the solutions, called eigenfunctions of the operator or wave functions, and $E$ stands for the associated energy eigenvalues.

Neglecting spin and relativistic effects the Hamilton operator of a molecule composed of $n$ electrons and $N$ atomic nuclei is

$$H = -\frac{\hbar}{2m}\sum_i^n \nabla_i^2 + \sum_{i,I}^{n,N} \frac{-e^2 Z_I}{4\pi\epsilon_0|\mathbf{r}_i - \mathbf{R}_I|} + \frac{1}{2}\sum_{i\neq j}^{n} \frac{e^2}{4\pi\epsilon_0|\mathbf{r}_i - \mathbf{r}_j|} \tag{2}$$

$$-\sum_I^N \frac{\hbar}{2M_I}\nabla_I^2 + \frac{1}{2}\sum_{I\neq J}^{N} \frac{Z_I Z_J e^2}{|\mathbf{R}_I - \mathbf{R}_J|}, \tag{3}$$

(a detailed discussion of this operator and the meaning of the various symbols is found in [23]). The resulting Schrödinger equation $H\Psi = E\Psi$ can be solved exactly for one electron and one nucleus. Exactly means that the associated eigenfunctions $\Psi$ are obtained as elementary functions that can be computed to any desired precision. In all other cases only approximate solutions of $E$ and $\Psi$ can be found. Here 'approximate' means two different things. On the one hand it means that the values of the solutions come out as a table of function arguments versus function values. And on the other hand this means that the respective function values are only approximations to the true but unknown values. The reliability of the results obtained is then usually judged by comparison with experimental data.

The solutions $\Psi$ have many different aspects. In the case of molecules consisting of more than one atom the nuclei are generally fixed in space, which is usually called the Born-Oppenheimer approximation, so that only the electrons need to be considered. For a specific configuration of the nuclei the positions of the electrons are then obtained from the scalar product of the wave functions $\Psi$.

The most interesting solution is the so called ground state, characterized by the lowest energy $E_0$ and the associated wave function $\Psi_0$. In chemical applications it usually suffices to know the ground state. But how do we know the positions of the nuclei? If we know the structure from experiment (e.g. X-ray), then, in principle, we can use the theory to confirm the structure. But in actual fact the argument usually goes the other way round. Since the solution of the Schrödinger equation for several atoms is a formidable computational challenge the experimental structure is usually used to judge the quality of the calculations. Deviations in the order of a few percentages of bond distances and bond angles between experimental structure and computed result are generally rated as an excellent agreement between theory and experiment.

But if we need to determine the unknown structure of a molecule by quantum mechanical calculations the only possibility is the systematic variation of the positions of the nuclei. For each configuration of the nuclei we then must find the associated ground state wave function for the electrons and the associated ground state energy. To find the most stable molecular structure we then have to find the minimum among all these ground state energies as a function of the positions of the nuclei. With the advent of density functional theory, and quantum Monte Carlo methods such computations have become feasible for small molecular systems, where small means a few atoms. The remarkable successes of these approaches have led to widespread interest in density functional theory as the most promising approach for accurate, practical methods in the theory of materials[23].

We must add however, that in the actual calculation of molecular systems we still have to deal with a series of nasty problems. Except for the hydrogen atom, we have to replace the actual Hamiltonian by a so called mean field approximation, so that we can treat the particles as independent. The reason for this is that presently we cannot properly handle the pairwise interactions of a system of more than two particles in our calculations. A large part of current quantum mechanical research is concerned with approaches that can be used to circumvent these obstacles.

As Paul Dirac noted in 1929:

> The underlying physical laws necessary for a mathematical theory of a large part of physics and the whole of chemistry are thus completely known, and the difficulty is only that the exact application of these laws leads to equations much too complicated to be soluble.

(quoted in [33]). Since then the computational side of quantum mechanics has made considerable progress, but with respect to larger molecules Dirac's assessment of 1929 is still up to date. There may be surprises ahead but so far the problems encountered in protein structure validation and prediction are not amenable to direct quantum mechanical computations. To make progress we have to try other approaches.

## 3. Molecular Mechanics Force Fields

Although direct quantum mechanical calculations are rarely applied to problems in protein structure theory some quantum mechanical results have found their way into protein structure prediction and refinement. The most popular model for pairwise atomic interactions is the Lennard-Jones potential

$$E(r) = -\frac{A}{r^6} + \frac{B}{r^{12}}. \tag{4}$$

This function describes the potential energy of the interaction of two noble gas atoms as a function of particle separation $r$. Most textbooks of biochemistry, physical chemistry, protein structure, and so on, explain the interactions found in proteins and other biological molecules in terms of this model and thus for many the Lennard-Jones potential has acquired the status of a physical law. To a large extent the Lennard-Jones potential owes its popularity to the simple functional form and the comprehensible physics embodied in this model. This is in stark contrast to the Copenhagen interpretation of quantum mechanical phenomena which according to Niels Bohr are necessarily incomprehensible to the human mind.

The attractive term, $r^6$, originates from the interaction of the two instantaneous dipole moments of the interacting atoms and the repulsive term, $r^{12}$, is due to the Coulomb repulsion of the core electrons. This energy function is a quantum mechanical result and its development has an interesting history[33]. By all standards the Lennard-Jones potential is a very good model for the interaction of two argon atoms. However, it turned out that for any other material than a noble gas this is a very poor approximation[10].

The Lennard-Jones potential is a major component of the standard empirical potential energy functions frequently employed in problems of protein structure research. The complete energy function is

$$\begin{aligned} E &= \sum K_b(b-b_0)^2 \\ &+ \sum K_\theta(\theta-\theta_0)^2 \\ &+ \sum K_\phi\left[1+\cos(n\phi-\delta)\right] \\ &+ \sum_{i<j} \frac{A_{ij}}{r_{ij}^{12}} - \frac{B_{ij}}{r_{ij}^{6}} \\ &+ \sum_{i<j} \frac{q_i q_j}{D_0 r_{ij}}. \end{aligned} \tag{5}$$

The first two terms model the lengths of covalent bonds and the angles between two consecutive covalent bonds, respectively. The third term describes the variation of energy as a function of the angle of a rotatable bond. The fourth term is the Lennard-Jones potential for pairs of atoms and the last term is the electrostatic interaction of atom pairs modeled by the Coulomb interaction, where $D_0$ is the dielectric constant.

It is generally assumed that this 'minimalist model'[8] captures the main components of the conformational energy of protein molecules, in the sense that it seems to be a satisfactory compromise between simplicity and accuracy[3]. In fact the model captures the local stereochemistry of protein structures and it is a very useful tool to remove steric clashes and local distortions in structural models. However, there are a number of problems. The parameters describing the specific properties of the various atom types are estimated from a multitude of experimental data and quantum mechanical considerations. These parameters are interdependent and errors in one parameter are propagated and perturb the assessment of other parameters (see e.g.[44]). But the most serious problem is the assumption that all non-bonded interactions can be modeled by a combination of a Lennard-Jones type potential and a Coulomb interaction. It has never been demonstrated that this assumption is justified.

To the contrary, Novotny, Bruccoleri and Karplus[30] have shown in 1984 that native and completely misfolded structures of proteins cannot be distinguished by energy calculations based on the model defined by equation 5. Subsequent studies reported improved performance when the aqueous environment is included in the calculations. Including water molecules in such calculations dramatically increases the complexity and computational demands. Among all problems involving molecular forces the problem of liquid water and its interactions with solutes stands out as the most challenging[33, 7] and Lennard-Jones potentials and Coulomb's law are inadequate models for these interactions. Adding water molecules in this scheme does not solve the problem.

# 4. Molecular Forces in Statistical Mechanics

Statistical mechanics provides the tools to investigate the configurations of a large number of particles and the associated molecular forces. One of the key results of classical statistical mechanics is Boltzmann's principle

$$p(x) = \frac{1}{Z} e^{-\epsilon(x)/kT}, \tag{6}$$

where

$$Z = \sum_{x} e^{-\epsilon(x)/kT}, \tag{7}$$

is called the Boltzmann sum or partition function. The principle states that the occupancy $p(x)$ of a certain state $x$ of a system is a direct consequence of the energy $\epsilon(x)$ of this state. If $x$ is the distance between two atoms of types $a$ and $b$ then $\epsilon_{a,b}(x)$ is the energy of interaction of these atoms at this distance and $p(x)$ is the associated probability density distribution that can be calculated from equation 6. The terms occupancy, density distribution and relative frequency are often used as synonyms for $p(x)$.

There are several routes that lead to Boltzmann's distribution. The approach frequently found in textbooks of statistical mechanics (e.g.[13]) is based on maximizing the number of microstates

$$\Omega = \frac{N!}{n_1! n_2! \cdots n_M!}, \qquad N = \sum_{i}^{M} n_i. \tag{8}$$

Here $N$ is usually interpreted as the number of particles and $M$ as the number of distinguishable states of the particles. We can be more general and interpret $N$ as the total number of cases and $M$ as the number of distinguishable cases. If we are interested in distances between atoms, for example, then $N$ may represent the number of distinct pairs of atom types $(a, b)$ found in our system and $M$ may refer to the number of distinct distance intervals used to record the separation of two atoms. Whatever states the indices $i$ may specify, the $n_i$ represent the number of cases that are found in state $i$ and we can express these occupancies also in terms of probabilities $p_i = n_i/N$. If our variables are, like distances, continuous we may use the indices to represent intervals. In this case we may write $p(x) = n(x)/n$. Note that there is no particular significance in using the symbols $x$, $i$, etc., since the index sets represented by these symbols are merely used to label the various states. A microstate is then defined by the exact specification of the particular state of each individual particle or case, whereas a macrostate is defined by the occupancies or frequencies $n_i$ or, equivalently, by the probabilities $p_i$, where $i = 1, \ldots, M$. In general, a particular macrostate can be realized by many distinct microstates.

A central assumption (often called postulate) of statistical mechanics is that the equilibrium state of a system corresponds to that particular macrostate which has the maximum number of microstates, $\Omega$, given the constraint of a constant number of cases, $\sum n_i = N = \text{constant}$, and given the constraint of constant energy $U$, i.e.

$$U = \sum_{i=1}^{M} n_i \epsilon_i = \text{constant} \tag{9}$$

The extremum of $\Omega$ is found by maximizing the logarithm of the number of microstates $\log \Omega$ by a variation of the absolute occupancies $n_i$ under the given constraints. This immediately yields Boltzmann's distribution (equation 6), where the temperature $T$ is obtained from the additional assumption (i.e. postulate) that the entropy $S$ is proportional to the logarithm of the number of microstates, $S = k \log \Omega$ ($k$ Boltzmann's constant), and the thermodynamic relationship $dS = dQ/T$, where $dQ$ is the heat input or output of a reversible thermodynamic process at temperature $T$.

To apply Boltzmann's principle we need to specify the energy function $\epsilon_i$ or $\epsilon(x)$ governing the system of interest. From the energy function and Boltzmann's principle we can then compute the occupancies of the various distinguishable states $n_i$ and the associated probabilities $p_i$.

Another route to the Boltzmann distribution (equation 6) is due to Williard Gibbs[12]. He starts out from the classical laws of motion of a system of particles. The motion is governed by the total energy, i.e. the sum of the kinetic and potential energy of the particles, which is called the Hamiltonian of the system. As it turns out there are some integrals of the motion that remain invariant if the system is in equilibrium. In a less technical language this means that some quantities remain constant even if the atoms or particles of the system are moving relative to each other. Two such quantities are the total energy and the probability density distribution expressed as a function of coordinates and momenta. This statement can be summarized in the following way

$$\sum_i \left( \frac{dP}{dq_i} \frac{d\epsilon}{dp_i} - \frac{dP}{dp_i} \frac{d\epsilon}{dq_i} \right) = 0, \tag{10}$$

where we have used the original notation of Williard Gibbs[12]. Here the probability density function $P = P(q_1, \ldots, p_1, \ldots)$ is a function of the coordinates $q_i$ and momenta $p_i$ and $\epsilon = \epsilon(q_1, \ldots, p_1, \ldots)$ is the energy of the system, which is again a function of coordinates and momenta. The sum extends over all particles $i = 1, \ldots N$. Gibbs then argues that a function of the form

$$P = e^{\frac{\Psi - \epsilon}{\Theta}} \tag{11}$$

seems to represent the most simple probability density function possible that satisfies equation 10. Here $\Theta$, called the modulus of the distribution, is later shown to be equivalent to the temperature, whereas $\Psi$ is a normalizing factor, defined by

$$e^{-\frac{\Psi}{\Theta}} = \int \cdots \int e^{-\frac{\epsilon}{\Theta}} dp_1 \cdots dq_n. \tag{12}$$

The integral is called partition function or Boltzmann sum which we have already encountered in its discrete form in equation 7.

Since the kinetic energy is a function of the momenta $p_i$ whereas the potential energy is a function of the coordinates $q_i$ it so happens that the partition function 'partitions' additively into a kinetic and a configurational part. This simplifies the calculations to some extent, since the equilibrium structure of a molecular system only depends on those variables needed to specify the structure of the system, like coordinates or distances between particles, while we may neglect the associated momenta.

The approach of Gibbs has the advantage that it is rooted in the most general version of classical mechanics, i.e. on the laws of motion expressed in Hamiltonian form. The main point is that the energy function of a system (i.e. the Hamiltonian) uniquely determines the probability density function $p(x)$. If the system is not disturbed by external forces, so that it remains in equilibrium, then the density distribution $p(x)$ is constant in time. To apply the theory we have to find or guess the energy function of the system from which we can deduce the density of states $p(x)$ and for this it does not matter whether we take the route of Boltzmann or that of Gibbs. But, as long as we do not know the functional form of the molecular interactions, $\epsilon(x)$, we cannot compute $p(x)$ and with respect to protein structure validation and prediction we are still at square one.

## 5. Molecular Forces and Radial Distribution Functions

A direct offspring of the principles of statistical mechanics is the theory of radial distribution functions and correlation functions which is the main subject of the theory of simple liquids[14, 5, 16]. This branch of statistical mechanics provides powerful tools for the investigation of molecular forces. The principal ideas of this theory are straightforward. Consider a system of particles or atoms. We may tag one of the atoms and determine the distribution of all other atoms as a function of the distance from the tagged atom. We may then repeat this analysis for all the remaining atoms and compute the average distribution. The result is a radial distribution function representing the average configuration of the particles in terms of inter-particle distances.

Radial distribution functions of liquid and solid materials can be determined by X-ray or neutron diffraction[14, 5]. These distribution functions can then be compared to theoretical results obtained from molecular simulations and other calculations. The essential relationship between molecular forces, interaction energies, and distribution functions is summarized in the following expression,

$$\overline{F}(r) = -\nabla\overline{U}(r) = kT\nabla \log[g(r)]. \tag{13}$$

Here $r$ is the distance between two particles and $F(r)$ is the force between two particles. The force is the negative gradient, denoted by $\nabla$, of the potential energy $U(r)$ between the two particles, and $g(r)$ is the radial distribution function. The bars over $F$ and $U$ remind us that these quantities are averages over all pairs of particles in the system. The symbol $g$ denoting the radial distribution function has no bar since it is explicitly defined as an average over many particles.

The detailed expression for the radial distribution function has the form of Boltzmann's law (equation 6)

$$g(r_{1,2}) = \frac{1}{Z}\int \cdots \int e^{-U(\mathbf{r})/kT} dr_3 \cdots dr_N \tag{14}$$

where

$$Z = \int \cdots \int e^{-U(\mathbf{r})/kT} d\mathbf{r}. \tag{15}$$

Here $\mathbf{r}$ denotes the positions of all particles $r_1, r_2, \ldots, r_N$. To obtain $g(r_{1,2})$ we integrate the Boltzmann factor, $\exp[-U(\mathbf{r})/kT]$, over all particles except two and normalize by the so called configuration integral, i.e. the configurational part of the partition function, which is just the integral over all possible positions of the particles. These expressions may appear somewhat unwieldy and as a matter of fact they are. A major reason for this is that in the derivation of these expressions we have to start from a notation that allows us to handle the positions of all particles although the final result depends on the distance between two particles.

We can summarize and interpret all this in a quite simple form. The radial distribution function can be written as

$$g(r) = \frac{n(r)}{n_g(r)}, \tag{16}$$

where $n(r)$ is the number density, i.e. the number of particles, found at distance $r$ from a central particle as observed in a material sample and $n_g(r)$ is the distribution of particles in an ideal gas having the same particle density as the sample. By definition, the particles in an ideal gas do not interact at all, i.e. there are no forces and the energy is the same for all distances, but the distribution contains all geometrical constraints which are impressed on the distribution of distances in a three dimensional space. Therefore, the ideal gas distribution $n_g(r)$ is a convenient reference state and the fraction, equation 16, measures the deviation of the distribution of distances in the material sample from the distribution of an ideal gas. For $g(r) > 1$ the interaction between two particles at distance $r$ is attractive and it is repulsive for $g(r) < 1$.

From the radial distribution function we obtain an energy function

$$w(r) = -kT \log[g(r)], \tag{17}$$

describing the average energy of interaction of two particles in the sample. Since this function is derived from the averaged force $\overline{F}(r)$ acting between all pairs of particles (equation 13) energy functions of this type are called potentials of average force and also potentials of mean force.

Radial distribution functions have been determined for a number of 'simple liquids', including water. Of course, water is not a simple liquid. The term 'simple' refers to the number of distinct atom types constituting the liquid and not to the complexity of the interactions involved. The functions obtained from diffraction experiments can then be compared to radial distribution functions obtained from molecular simulations based on Boltzmann's distribution (equations 6 and 16). There is an extensive literature on simulations of systems of hard spheres and systems of particles that interact according to Lennard-Jones type potentials[14, 5]. In general the basic shapes of the functions derived from such simulations agree at least qualitatively with the radial distribution functions obtained from experiment.

In general, radial distribution functions can be obtained only for simple molecular systems. The reason is that radial distribution functions measured on liquid or solid materials consisting of more than one atom type are combinations or superpositions of distinct atom pair radial distribution functions. For example, the radial distribution function of water is a superposition of three types of functions corresponding to the interactions of a pair of hydrogen atoms, a pair of oxygen atoms, and a pair of one hydrogen and one oxygen atom and it is not an easy matter to extract these individual components from the total radial distribution function.

Radial distribution functions and potentials of mean force emphasize the direct link between molecular forces and molecular structure. Of course, this connection is already implicitly contained in Boltzmann's law, equation 6, but the theory of radial distribution functions provides the appropriate mathematical tools to accomplish the transformations between energy and structure and it yields a plausible interpretation of the physical principles involved. The connection between energy and structure provides the means for the quantitative analysis of the forces and interactions in proteins as we will see shortly. Before we embark on this endeavor it is instructive to briefly contemplate the conditions under which physics was practiced when the foundations of statistical mechanics and quantum mechanics were framed and elaborated.

## 6. Molecular Forces and Physical Theories

The fundamental physical theories of thermodynamics, statistical mechanics, quantum mechanics, relativity, etc., and the associated mathematical tools have been invented within a period of a hundred years, which extends roughly from 1850-1950. The physicists of these days generally had a small number of fundamental experimental data and the physical theories were invented to explain the experimental facts. If this failed, attempts were made to find new basic principles. The most dramatic example is certainly the development of quantum mechanics. Starting from a few basic assumptions the theory developed between 1900 and 1930 eventually managed to comprehend and 'explain' a large body of important experimental facts, like black body radiation, atomic spectra, chemical bonding, etc., with unprecedented precision.

Starting with the second half of the twentieth century the data processing power of electronic computers opened possibilities that were unthinkable in the days of Boltzmann, Schrödinger, or Dirac. Detailed calculations involving a fistful of particles are impractical without a computer and limited capabilities in data processing and numerical work necessarily constrain the possible approaches to a physical problem. These limitations shaped the very definition of physics and what was considered a tractable physical problem. Even in the opinion of Boltzmann, proponent of the atomic theory of matter and inventor of the statistical theory of gases, the computation of distances and forces for a large number of individual atoms was not an attractive option. In 1895 Boltzmann remarks,

> the difficulty of enumerating all the material points of the universe and of determining the law of mutual force for each pair, would only be a quantitative one; nature would be a difficult problem, but not a mystery for the human mind.

(quoted in[33]). The meaning of Boltzmann's prose is not entirely clear but he seems to say that a detailed description of a large number of specific molecular configurations and interactions is a boring venture and does not yield to general physical laws. We must concede to Boltzmann that he and his contemporaries never had the chance to see the atomic structure of a protein and in his time the existence of such complex but nonetheless well defined molecular structures must have been unthinkable. Today we have to face this complexity and to make progress we have to determine the laws of mutual force of all the various types of atomic interactions found in biological structures, may their number be large or small.

For this venture the methods of statistical mechanics, laid out by Boltzmann, Gibbs and many others, provide an appropriate basis. To proceed we emphasize one additional point regarding Boltzmann's law, equation 6. As it stands, the expression implies that the probability density, $p(x)$, is a function of the energy $\epsilon(x)$. Once we know the energy function $\epsilon(x)$, we are able to compute $p(x)$ and hence the structure of the system. And from the associated partition function, equation 7, we can derive thermodynamic quantities like entropy, Helmholtz free energy, and so on. An approach like this is a characteristic example of physical reasoning. In the language of physics, a true understanding and explanation of a natural phenomenon is achieved, if we can derive the energy function or the energy operator from first principles and if we can show that the laws of quantum mechanics, classical mechanics, and statistical mechanics when applied to this energy function reproduce the experimental facts to some degree of accuracy.

With this in mind we understand Boltzmann's reasoning that a proper description of atomic interactions requires that we can reduce the problem to a small set of basic laws of force. A large number of distinct energy functions, that need to be adapted to each individual case, is of little value. We also see that the energy operator, equation 3, used to express the physics of a molecular system in quantum mechanical terms, contains a simple electrostatic energy function and thus satisfies Boltzmann's demand. Neglecting spin, all the interactions among electrons and nuclei are governed by Coulomb's law, where the energy function has the most simple form $q_a q_b / r$, where the $q$'s correspond to the charges carried by electrons and nuclei (i.e. the $e$ and $Z$ in equation 3), and where $r$ is the separation of the particles. All the miracles of chemistry and biology seem to follow from this disarmingly simple function. But as we have already seen, the structures of proteins are beyond the present capabilities of this astonishing theory.

## 7. Molecular Forces and Molecular Structure

The energy function plays a central role in the description of a physical system and most other quantities, in particular density distribution functions, are considered to be quantities of less nobility. Inverting Boltzmann's law (equation 6),

$$\epsilon(x) = -kT \log p(x) - kT \log Z \tag{18}$$

the energy $\epsilon(x)$ is obtained from the density distribution $p(x)$. When applied to molecular interactions the expression shows that the 'laws of force' can be derived from an experimental determination of the density distribution $p(x)$ of the configurational variables $x$, where the experiments are carried out at a certain temperature $T$. Here the energy has lost its dominant position and the two versions of Boltzmann's law show that in actual fact energy and density are on an equal footing and they are two sides of the same coin.

To apply the inverted form of Boltzmann's law in the analysis of atomic interactions we need a proper amount of highly resolved and reliable experimental data on the relative arrangements of atoms so that we can compile density distributions of sufficient accuracy. The amount of detailed structural information on proteins and on other molecular systems acquired over the last decades is enormous and suffices to compile potentials of average force for most interactions encountered in the structures of proteins.

We briefly review the steps required to compile potentials of mean force from a set of protein structures[37, 38]. We write Boltzmann's law in the form

$$E(a,b,r) = -kT \log g(a,b,r) \tag{19}$$

where $E(a,b,r)$ is the generic potential of mean force for the interaction of two atoms of type $a$ and $b$ at separation $r$, $g(a,b,r)$ is the generic two particle radial distribution function, and $k$ is Boltzmann's constant and $T$ is the absolute temperature. The potential function $E(a,b,r)$ may be split into an annealed and a quenched part. The annealed or 'soft' part is specific for the particular atom types $a$ and $b$. The quenched or 'hard' contributions are intrinsic to the covalent structure and compact nature of protein structures and they also contain the subtle constraints imposed by the geometry of three dimensional space. The quenched contributions are extracted from the generic radial distribution functions by averaging over all atom pairs $a$ and $b$ which yields the unspecific radial distribution function

$$g(r) = \frac{1}{n} \sum_{a,b} g(a,b,r) \tag{20}$$

where $n$ is the number of distinct generic distribution functions. The specific radial distribution functions are defined as the fractions

$$g_s(a,b,r) = \frac{g(a,b,r)}{g(r)}. \tag{21}$$

Here, in analogy to the ideal gas distribution of equation 16, the unspecific radial distribution function acts as a reference state. From the specific radial distribution functions $g_s(a,b,r)$ we obtain the specific potential of mean force

$$\epsilon(a,b,r) = -kT \log \frac{g(a,b,r)}{g(r)} = -kT \log g(a,b,r) + kT \log g(r). \tag{22}$$

In all these expressions we have implicitly used the convenient choice $-kT \log Z = 0$, but we could have used any other finite value since with respect to the potential of mean force at constant temperature $T$ this term is an additive constant.

Equation 22 summarizes the basic recipe for the compilation of a set of potentials of mean force. The compilation of radial distribution functions and the application of equation 22 to the atom types found in protein structures requires a number of finer details that need to be considered. For example, some atom types are rare so that radial distribution functions compiled for these atom types may have rather large fluctuations as compared to the more frequent atom types. To handle this and other problems proper techniques are available[37].

## 8. Molecular Forces and Protein Structures

Work on mean force potentials for the interactions found in proteins has started in 1990[37] and it was quickly demonstrated that these functions can be used to distinguish native from misfolded structures[37, 15] and several erroneous experimental structures were detected in the public domain[39, 40]. Concurrently the efficiency of mean force calculations enabled the development of new prediction techniques like fold recognition and threading[46, 18, 21, 42], and techniques like fragment assembly were used to correctly predict approximate structures of small proteins ahead of experiment[43]. In the following years mean force potentials were used to calculate the changes in stability associated with amino acid mutations[40], and the were used in protein design[51] and the docking and binding of small molecules to proteins[29]. An example of applications to protein sequence randomization and structural stability is summarized in figure 4. Early implementations of mean force potentials use simplified representations of protein structures considering only the $C^\alpha$ and $C^\beta$ atoms along the polypeptide chain. One advantage of such reduced models is their computational efficiency enabling the search of large sequence and structure data bases and the generation and evaluation of a large number of conformations to find structures that are as close as possible to the native fold of a protein.

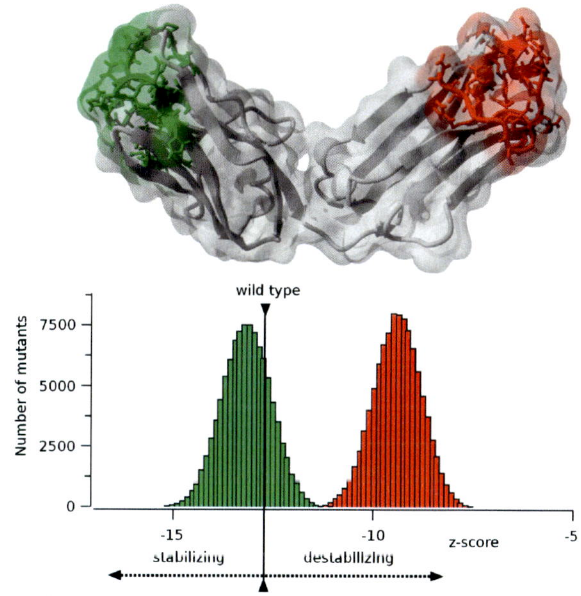


Figure 4. The stability of antibody mutants investigated by potentials of mean force. The complementarity determining regions (CDR) of an antibody are shown in green and an arbitrary region of similar architecture (a 'control' region) is shown in red. The histogram shows the distribution of scores of randomized sequences computed from mean force potentials. More than 70% of the $10^5$ randomized CDR sequences stabilize the immunoglobulin fold (green). In contrast, all randomized sequences in the control region destabilize the fold (red) [51].

It is clear however, that the accurate validation and reliable correction of protein structures and the prediction of structures with a precision comparable to experimental structure determination requires force fields containing all the interactions found in proteins. The development of a complete set of mean force potentials[41, 45, 35] is the subject of current research[50, 49]. To illustrate the current state of the art we return to our introductory example of asparagine and glutamine rotamers.

The amide groups of asparagine and glutamine are specific examples of functional groups which act simultaneously as hydrogen bond donors and acceptors. Hydrogen bonds form when a strongly electronegative nitrogen or oxygen atom, the hydrogen bond donor, shares a covalently attached proton with a lone pair of electrons of another oxygen atom, called the hydrogen bond acceptor. In proteins hydrogen bond donors and acceptors are always an integral part of the polypeptide backbone or the amino acid side chains. Since molecular forces strongly depend on the covalent structure and chemical environment in the vicinity of the interacting atoms there are many individual types of hydrogen bonds in proteins.

The variability of hydrogen bonds and other interactions is easily captured by mean force potentials. The only requirement is that we distinguish among the various atom types and compile potentials for the individual interactions. In particular the backbone atoms N, $C^{\alpha}$, C', and O of the various amino acids are distinct atom types. With this distinction the standard amino acid residues found in protein crystal structures determined by X-ray analysis contain $n = 167$ distinct atom types resulting in a total of $(n + 1) \times n/2 = 14,028$ interactions. This number does not include any explicit interactions involving hydrogen atoms, atoms of non-standard groups, or water. But these interactions are not completely neglected. Mean force potentials capture contributions from the complete chemical environment even if the latter is not completely specified. Therefore, an immediate advantage of potentials of mean force is that the evaluation of interaction energies between hydrogen bond donors and acceptors does not require the explicit consideration of hydrogen atoms.

However, it is known in advance that the data base contains a substantial fraction of incorrect asparagine and glutamine rotamers and at the outset it is unclear to what extent the error rate of more than 20% corrupts the radial distribution functions resulting in defective or unusable mean force potentials. The compilation of mean force potentials from defective data is in fact an intriguing challenge.

As it turns out the effect of errors on the potentials is quite small. Moreover, a refinement cycle consisting of rotamer correction and recompilation of potentials quickly converges to a stable solution and there is excellent agreement between the rotamer flips suggested by a thorough analysis based on physico-chemical principles[52] and those suggested by mean force calculations. Figures 1, 2, and 3 provide examples of incorrect rotamers as found in highly resolved protein crystal structures and the respective corrected configurations. Examples of potentials before and after refinement are shown in figure 5.

Potentials of mean force have several remarkable properties and difficult problems of protein structure analysis and prediction can be tackled quite successfully. The reason for this is that mean force potentials provide a concise and compact representation of the complete experimental knowledge on protein structures currently available. The only assumption or 'postulate' is that the relationship between structure and energy, as expressed by Boltzmann's law (equation 22) holds true. There are no additional physical parameters that enter the calculations, like the $A$'s and $B$'s of Lennard-Jones potentials, dielectric constants, partial charges, and so on, required to implement a minimal version of molecular force fields. Similarly, energy calculations based on potentials of mean force are straightforward and efficient and there is no need for sophisticated mathematical procedures as they are required for the solution of quantum mechanical differential equations. A most attrac-

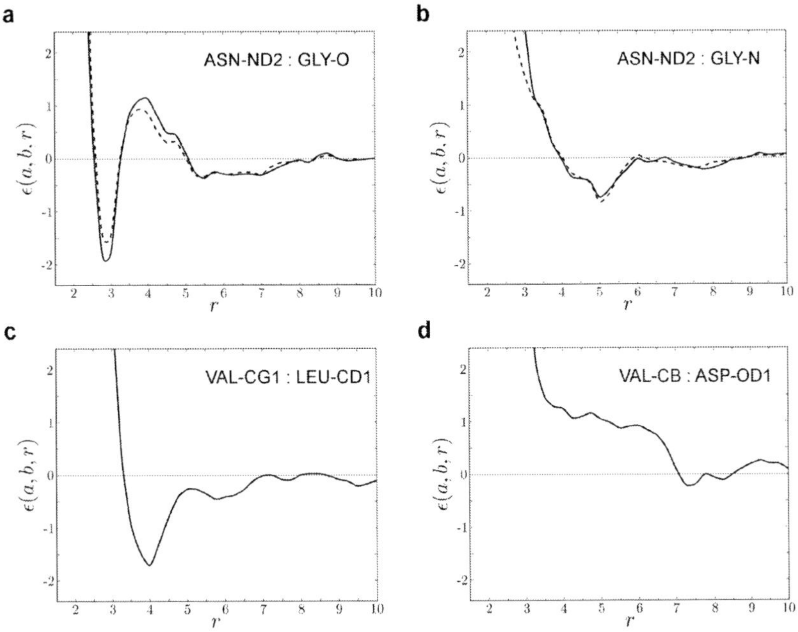


Figure 5. Examples of mean force potentials. Refined (solid lines) and unrefined (dashed lines) potentials of mean force, $\epsilon(a, b, r)$, between atoms of type $a$ and $b$ are plotted as a function of separation $r$ in the distance range $1.5 \leq r \leq 10.0$ Å. All potentials converge to zero at separations larger than $r \approx 15$ Å. The atom types shown are asparagine side-chain amide nitrogen (ASN-ND2), glycine backbone oxygen (GLY-O), glycine backbone nitrogen (GLY-N), valine terminal side-chain carbon (VAL-CG1), leucine terminal side-chain carbon (LEU-CD1), valine $\beta$ carbon (VAL-CB) and aspartic acid terminal side-chain oxygen (ASP-OD1). **a,** Interaction of the hydrogen bond donor ASN-ND2 and hydrogen bond acceptor GLY-O. **b,** Interaction of the hydrogen bond donor ASN-ND2 and hydrogen bond donor GLY N. **c,** Interaction of the aliphatic carbon atom VAL-CG1 and the aliphatic carbon atom LEU-CD1. **d,** Interaction of the aliphatic carbon atom VAL-CB and the charged oxygen atom ASP-OD1.

tive property of mean force potentials is that they are self-consistent and self-correcting. Errors and inconsistencies in the data from which mean force potentials are compiled are detected by the very same potentials[50].

## 9. Molecular Forces of Hydrogen Bonds

Figure 6 shows the 'law of force' for four types of interactions found in protein structures. Each interaction is represented by two functions. The radial distribution functions are compiled from a data base of 833 highly resolved protein crystal structures. The solid lines correspond to potentials of mean force obtained from single protein chains and the respective radial distribution functions solely contain intramolecular distances. The dashed lines correspond to mean force potentials compiled from complete protein crystal structures. Here the radial distribution functions contain inter- as well as intramolecular distances. Nevertheless the two types of functions are quite similar demonstrating that intra- and intermolecular interactions follow the same basic rules and it also shows that the functional form of the individual potentials are quite stable with respect to variations in the data sources. Similar results are obtained when data sets are split into independent subsets. A comparison of independent data sets generally reveals that corresponding potentials are practically indistinguishable.

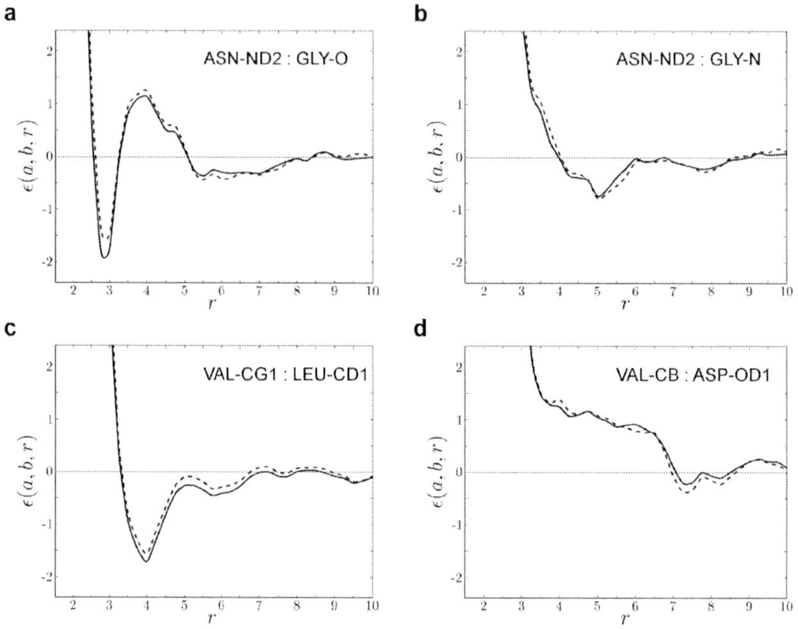


Figure 6. Comparison of mean force potentials compiled from single molecules (dashed lines) and complete crystal structures (solid lines). The atom types shown correspond to those of figure 5.

The interactions found in proteins are often grouped in polar, ionic, hydrophobic and hydrogen bond interactions. Figures 5 and 6 show a hydrogen bond interaction (a), a polar interaction (b), a hydrophobic interaction (c), and an interaction between hydrophobic and a polar groups (d). The interaction between the $C^{\gamma}$ group of valine and the $C^{\delta}$ group of leucine (c), a typical hydrophobic interaction, has a deep minimum and is attractive over the whole distance range except for short distances where the atoms start to penetrate each other. The interaction between the hydrophobic $C^{\beta}$ atom of valine and the the charged $O^{\delta}$ atom of aspartic acid (d) is predominantly repulsive. The interaction between the asparagine $N^{\delta_2}$ atom and the glycine backbone nitrogen (b) has an appreciable attractive energy well. This potential describes the interaction of two positive partial charges and one might expect that the interaction should be repulsive. However, the minimum is at the rather large distance of 5 Å. The attraction of these two atoms is mediated by a water molecule that is sandwiched between the two nitrogen atoms or some other group that can compensate the partial charges on the nitrogen atoms. This example demonstrates that the potentials encode the locations of intervening solvent molecules although the positions of these molecules are not explicitly specified.

The interaction between the $N^{\delta_2}$ atom of asparagine and the backbone oxygen of glycine (a) is a typical example of a hydrogen bond interaction. The potential has a deep narrow minimum at hydrogen bond contact corresponding to a short distance of 2.7 Å, separated by a high energy barrier from larger distances. Quite generally, mean force hydrogen bond interactions have the form of a molecular lock or shutter, characterized by an energy barrier with a maximum at $r_m$, separating a narrow energy well at hydrogen bond contact at $r_c$ from large distances, as shown in figure 7.

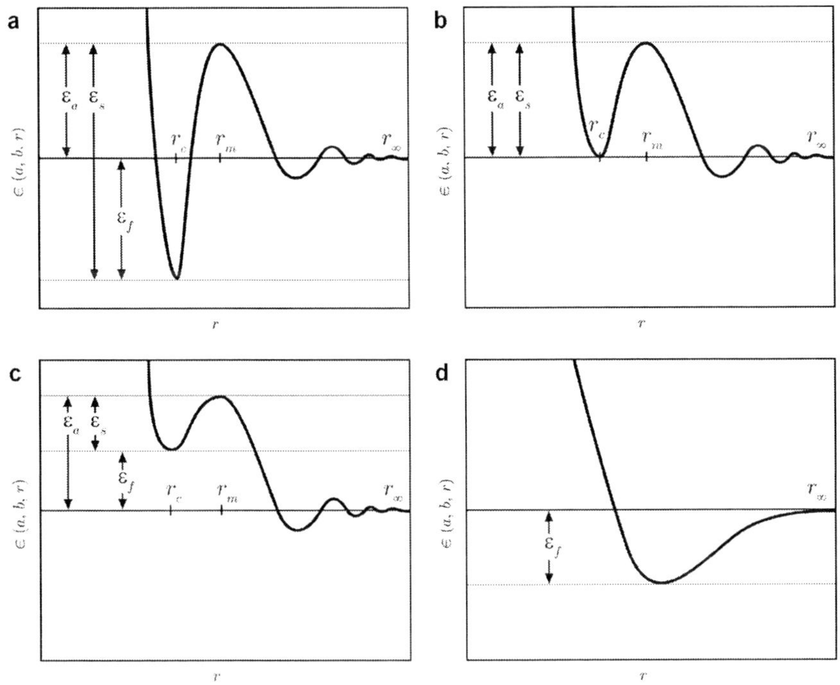


Figure 7. Generic functional forms of mean force potentials found in protein crystal structures. The generic forms of mean force potentials can be characterized by the free energy balance of bond formation, $\epsilon_f = \epsilon(r_c) - \epsilon(r_\infty) = \epsilon(r_c)$, the activation free energy, $\epsilon_a = \epsilon(r_m)$, required to surmount the barrier, and the activation energy of bond disruption, $\epsilon_s = \epsilon_f + \epsilon_a$, required to break the bond. **a,** Molecular lock corresponding to the potential shown in Figure 5 (a). Since $\epsilon_f < 0$, energy is released in the overall process of bond formation. **b,** Special case of a lock where $\epsilon_r = 0$ so that the energy balance of bond formation is zero. **c,** Molecular lock which requires energy input for bond formation (Figure 3 of[50] shows actual examples of this type of interaction). Here bond formation consumes free energy ($\epsilon_f > 0$). **d,** Typical interaction without a barrier corresponding to the interaction of aliphatic atoms of Figure 5 (c). Hydrogen bonds and other highly polar interactions generally have one of the functional forms (a-c), whereas (d) is the typical functional form for hydrophobic interactions.

The molecular lock is a general model for spatially precise and kinetically stable interactions where precision is determined by the width of the energy well and stability is mediated by the hight of the energy barrier, $\epsilon(r_m) - \epsilon(r_c)$. A consequence of this model is that stable bonds can be formed even if the energy balance of bond formation is positive, zero, or only slightly negative and all these types of hydrogen bonds are actually observed in protein structures[50]. The barrier is a specific feature of hydrogen bonds and highly polar interactions. Other interactions, in particular hydrophobic 'bonds', have comparatively broad minima and they lack barriers. Such interactions are less precise and their stability is determined by $\epsilon_f$ the depth of the energy minimum as shown in figure 7 (d).

Proteins have a large number of stabilizing hydrogen bond interactions although the free energy differences between folded and unfolded states is comparatively small. A generally accepted qualitative explanation for this apparent paradox is that in the folding of proteins intermolecular hydrogen bonds between protein atoms and water molecules are replaced by intramolecular hydrogen bonds between protein atoms on the one hand and hydrogen bonds between water molecules on the other, where the liberation of water molecules from the protein compensates for the loss in entropy of the 'freezing' protein. The individual contributions of the various energy terms are large but they almost cancel so that the resulting total energy difference between folded and unfolded proteins is small. Failures to compute the energy difference of protein folding from theoretical models are generally attributed to uncertainties arising from the cancelation of large energy terms.

In contrast, a resolution of this paradox in terms of mean force potentials is straightforward. Hydrogen bonds are molecular locks that stabilize protein structures whether the free energy balance of bond formation is negative, positive, or zero. Even if the total sum of all interactions is comparatively small the structure of a protein can be quite stable since the energy barriers prevent unfolding. Nevertheless, a final resolution of such paradoxes and other miracles of protein folding requires that we are able to compute protein structures with a precision and confidence that rivals experimental structure determination and that we can reproduce the free energies, heat capacities, etc., obtained in the laboratory. After all, this is still a considerable challenge.

## 10. Bibliographical Notes

The amount of available literature on molecular forces, the theory of protein structure, and protein structure prediction is enormous. Our aim here is to provide a set of general pointers to the literature and to extend the references mentioned in the text.

The history of molecular forces from Newton to the present day is laid out in a recent book by Rowlinson[33]. There are many excellent texts on quantum mechanics. The basic concepts discussed here can be found in the recent monograph of Martin[23] (where equation 3 is discussed in some detail) or Atkins and Friedman[1]. Both books provide the technical details required for quantum chemical computations. The books by Cramer[6] and Thijssen[48] are specifically dedicated to computational work in quantum mechanics. A recent text on the quantum mechanics of intermolecular forces is the book by Finnis[10].

Empirical force fields and their applications in molecular simulations are extensively covered by Leach[20] and others[3]. A recent update of force field parameters is found in[19]. The history of the Lennard-Jones potential is extensively covered by Rowlinson[33] and in a more technical form by Finnis[10]. There are numerous text books covering classical statistical mechanics. Brief accounts of the basic principles are found in [5, 13] and it is certainly advisable to consult Gibbs' elementary principles of statistical mechanics[12]. An accessible account of the theory of radial distribution functions and liquids is provided by Chandlers book[5] whereas the monograph by Hansen and McDonald[14] is a detailed account of the subject. A classic text in this area is the book by Hill[16]. The statistical mechanics of chain molecules, highly relevant to protein structure theory, is now out of fashion. The books by Flory[11] and Rubinstein and Colby[34] are excellent introductions.

Mean force potentials for proteins are introduced in[37] and reviewed in[38, 40]. A recent publication emphasizing the informational theoretical aspects is provided by Solis and Rackovsky[47]. Examples of various applications and extensions of mean force potentials are found in[36, 32, 22, 24, 4, 53, 25].

An introduction to the main aspects of protein structure theory and protein structure prediction is found in the book of Finkelstein and Ptitsyn[9]. A recent review on fold recognition and related techniques is provided by Mizuguchi[26]. Examples of recent reviews on various aspects of protein structure prediction are [31, 27, 2, 17] and the entry point to the most recent CASP experiment is[28].

## References

[1] P. W. Atkins and R. S. Friedman. *Molecular Quantum Mechanics.* Oxford University Press, 3 edition, 1997.

[2] Philip Bradley, Kira M S Misura, and David Baker. Toward high-resolution de novo structure prediction for small proteins. *Science*, **309**(5742):1868–1871, Sep 2005.

[3] Charles Brooks, Martin Karplus, and Montgomery Pettitt. *Proteins. A Theoretical Perspective of Dynamics, Structure, and Thermodynamics*. John Wiley & Sons, 1988.

[4] Nicolae-Viorel Buchete, John E Straub, and Devarajan Thirumalai. Orientational potentials extracted from protein structures improve native fold recognition. *Protein Sci*, **13**(4):862–874, Apr 2004.

[5] David Chandler. *Introduction to Modern Statistical Mechanics*. Oxford University Press, 1987.

[6] Christopher J. Cramer. *Essentials of Computational Chemistry.* Wiley, 2002.

[7] David Eisenberg and Walter Kauzmann. *The Structure and Properties of Water.* Oxford University Press, 1969.

[8] Alan Fersht. *Structure and Mechanism in Protein Science: A Guide to Enzyme Catalysis and Protein Folding*. W H Freeman, 1999.

[9] Alexei V. Finkelstein and Oleg B. Ptitsyn. *Protein Physics*. Academic Press, 2002.

[10] Mike Finnis. *Interatomic Forces in Condensed Matter*. Oxford University Press, 2004.

[11] Paul J. Flory. *Statistical Mechanics of Chain Molecules.* Wiley, 1969.

[12] J. Williard Gibbs. *Elementary Principles of Statistical Mechanics.* Ox Bow Press, 1901.

[13] A. M. Glazer and J. S. Wark. *Statistical Mechanics. A Survival Guide.* Oxford University Press, 2001.

[14] Jean-Pierre Hansen and Ian R. McDonald. *Theory of Simple Liquids.* Elsevier, 2006.

[15] M. Hendlich, P. Lackner, S. Weitckus, H. Floeckner, R. Froschauer, K. Gottsbacher, G. Casari, and M. J. Sippl. Identification of native protein folds amongst a large number of incorrect models. The calculation of low energy conformations from potentials of mean force. *J Mol Biol*, **216**(1):167 180, Nov 1990.

[16] Terrell L. Hill. *Statistical Mechanics. Principles and Selected Applications.* McGraw-Hill, 1956.

[17] Jol Janin and Michael Levitt. Theory and simulation Accuracy and reliability in modelling proteins and complexes. *Curr Opin Struct Biol*, **16**(2):139–141, Apr 2006.

[18] D. T. Jones, W. R. Taylor, and J. M. Thornton. A new approach to protein fold recognition. *Nature*, **358**(6381):86–89, Jul 1992.

[19] Elmar Krieger, Tom Darden, Sander B Nabuurs, Alexei Finkelstein, and Gert Vriend. Making optimal use of empirical energy functions: force-field parameterization in crystal space. *Proteins*, **57**(4):678–683, Dec 2004.

[20] Andrew R. Leach. *Molecular Modelling. Principles and Applications.* Prentice Hall, 2 edition, 2001.

[21] R. Lthy, J. U. Bowie, and D. Eisenberg. Assessment of protein models with three-dimensional profiles. *Nature*, **356**(6364):83–85, Mar 1992.

[22] H. Lu and J. Skolnick. A distance-dependent atomic knowledge-based potential for improved protein structure selection. *Proteins*, **44**(3):223–232, Aug 2001.

[23] Richard M. Martin. *Electronic Structure*. Cambridge University Press, 2004.

[24] F. Melo, R. Sanchez, and A. Sali. Statistical potentials for fold assessment. *Protein Sci*, **11**(2):430–448, 2002.

[25] Sanzo Miyazawa and Robert L Jernigan. How effective for fold recognition is a potential of mean force that includes relative orientations between contacting residues in proteins? *J Chem Phys*, **122**(2):024901, Jan 2005.

[26] Kenji Mizuguchi. Fold recognition for drug discovery. *Drug Discovery Today*, **3**(1):18–23, February 2004.

[27] John Moult. A decade of CASP: progress, bottlenecks and prognosis in protein structure prediction. *Curr Opin Struct Biol*, **15**(3):285–289, Jun 2005.

[28] John Moult, Krzysztof Fidelis, Burkhard Rost, Tim Hubbard, and Anna Tramontano. Critical assessment of methods of protein structure prediction (CASP)-Round 6. *Proteins*, **61** Suppl 7:3–7, 2005.

[29] I. Muegge and Y. C. Martin. A general and fast scoring function for protein-ligand interactions: a simplified potential approach. *J Med Chem*, **42**(5):791–804, Mar 1999.

[30] J. Novotny, R. Bruccoleri, and M. Karplus. An analysis of incorrectly folded protein models. Implications for structure predictions. *J Mol Biol*, **177**:787–818, 1984.

[31] Donald Petrey and Barry Honig. Protein structure prediction: inroads to biology. *Mol Cell*, **20**(6):811–819, Dec 2005.

[32] B. A. Reva, J. Skolnick, and A. V. Finkelstein. Averaging interaction energies over homologs improves protein fold recognition in gapless threading. *Proteins*, **35**(3):353–359, 1999.

[33] John Rowlinson. *Cohesion. A Scientific History of Intermolecular Forces.* Cambridge University Press, 2002.

[34] Michael Rubinstein and Ralph H. Colby. *Polymer Physics.* Oxford University Press, 2003.

[35] R. Samudrala and J. Moult. An all-atom distance-dependent conditional probability discriminatory function for protein structure prediction. *J Mol Biol*, **275**(5):895–916, 1998.

[36] K. T. Simons, C. Kooperberg, E. Huang, and D. Baker. Assembly of protein tertiary structures from fragments with similar local sequences using simulated annealing and bayesian scoring functions. *J Mol Biol*, **268**(1):209–225, 1997.

[37] M. J. Sippl. Calculation of conformational ensembles from potentials of mean force. An approach to the knowledge-based prediction of local structures in globular proteins. *J Mol Biol*, **213**(4):859–883, Jun 1990.

[38] M. J. Sippl. Boltzmann's principle, knowledge-based mean fields and protein folding. An approach to the computational determination of protein structures. *J Comput Aided Mol Des*, **7**(4):473–501, Aug 1993.

[39] M. J. Sippl. Recognition of errors in three-dimensional structures of proteins. *Proteins*, **17**(4):355–362, 1993.

[40] M. J. Sippl. Knowledge-based potentials for proteins. *Curr Opin Struct Biol*, **5**:229–235, 1995.

[41] M. J. Sippl. Helmholtz free energy of peptide hydrogen bonds in proteins. *J Mol Biol*, **260**:644–648, 1996.

[42] M. J. Sippl and H. Flockner. Threading thrills and threats. *Structure*, **4**:15–19, 1996.

[43] M. J. Sippl, M. Hendlich, and P. Lackner. Assembly of polypeptide and protein backbone conformations from low energy ensembles of short fragments: Development of strategies and construction of models for myoglobin, lysozyme, and thymosin beta 4. *Protein Sci.*, **1**:625–640, 1992.

[44] M. J. Sippl, G. Nemethy, and H. A. Scheraga. Intermolecular potentials from crystal data. 6. determination of empirical potentials for O-H...O=C hydrogen bonds from packing configurations. *J Phys Chem*, **88**:6231–6233, 1984.

[45] M. J. Sippl, M. Ortner, M. Jaritz, P. Lackner, and H. Flockner. Helmholtz free energies of atom pair interactions in proteins. *Fold Des*, **1**:289–298, 1996.

[46] M. J. Sippl and S. Weitckus. Detection of native-like models for amino acid sequences of unknown three-dimensional structure in a data base of known protein conformations. *Proteins*, **13**(3):258–271, Jul 1992.

[47] Armando D Solis and S. Rackovsky. Improvement of statistical potentials and threading score functions using information maximization. *Proteins*, **62**(4):892–908, Mar 2006.

[48] J. M. Thijssen. *Computational Physics*. Cambridge University Press, 1999.

[49] Christian X. Weichenberger and Manfred J. Sippl. NQ-Flipper: Validation and correction of asparagine/glutamine amide rotamers in protein crystal structures. *Bioinformatics (in press)*, 2006.

[50] Christian X. Weichenberger and Manfred J. Sippl. Self-consistent assignment of asparagine and glutamine amide rotamers in protein crystal structures. *Structure (in press)*, 2006.

[51] Markus Wiederstein and Manfred J Sippl. Protein sequence randomization: efficient estimation of protein stability using knowledge-based potentials. *J Mol Biol*, **345**(5):1199–1212, Feb 2005.

[52] J. M. Word, S. C. Lovell, J. S. Richardson, and D. C. Richardson. Asparagine and glutamine: using hydrogen atom contacts in the choice of side-chain amide orientation. *J Mol Biol*, **285**(4):1735–1747, Jan 1999.

[53] Chi Zhang, Song Liu, Hongyi Zhou, and Yaoqi Zhou. An accurate, residue-level, pair potential of mean force for folding and binding based on the distance-scaled, ideal-gas reference state. *Protein Sci*, **13**(2):400–411, Feb 2004.

In: Protein Folding-Misfolding...
Editors: J.P. Zbilut and T. Scheibel, pp. 141-162
ISBN: 1-60021-417-7


*Chapter 6*

# RECURRENCE QUANTIFICATION AS A SIGNAL ANALYSIS APPROACH TO THE STUDY OF PROTEIN SEQUENCE AND STRUCTURE

***Alessandro Giuliani[1*] and Joseph P. Zbilut[2**]***
Istituto Superiore di Sanita', Dipartimento Ambiente e Connessa Prevenzione Primaria
Viale Regina Elena 299, 00161 Roma, Italia
Department of Molecular Biophysics and Physiology, Rush University Medical Center
1653 W. Congress, Chicago, IL 60612 USA

## Abstract

Recurrence quantification analysis (RQA), is a signal analysis method to study sequence/structure relationships of proteins, a technque derived from nonlinear dynamics, which exposes and quantifies patterns in higher dimensional spaces. The basic focus of applications is in generating predictive sequence/structure or sequence/activity models on the basis of RQA hydrophobicity sequences of various protein families.

## Introduction

The relationship between sequence-embedded information and folding behavior of proteins is currently a dominant concern of both theoretical and applied biochemical research. More specifically, this concern redounds to areas such as (a) sequence-based functional predictions, (b) 3D structure-based functional predictions, (c) folding mechanism elucidation, (d) protein-protein interaction prediction and modelling [1,2].

From a purely information-theoretical point of view this problem can be expressed in terms of how a mono-dimensional string of amino acid residues ends up in a three-dimensional arrangement corresponding to protein three-dimensional structure in solution [2]. It is important to stress that "three-dimensional structure" should not be intended as a fixed

[*] E-mail address: alessandro.giuliani@iss.it
[**] E-mail address: jzbilut@rush.edu

architecture: many proteins appear as partially or even totally disordered when analysed with spectroscopic methods [3]; however, this apparent disorder corresponds to an efficient organization for protein physiological function. Thus we must attach to the word structure, when in presence of proteins, a mixed architectural and dynamical meaning.

How does the translation from mono-dimensional to three-dimensional take place? Who does the interpretation of information embedded in protein primary structures (the juxtaposition of residues) translating it into a meaningful three dimensional pattern? There is no "external reader" attaching a 3D translation to mono-dimensional sequences of amino acids; thus, the sequences must "embed" their own code that in turn attains its "correct" translation when in presence of the relevant chemico-physical environment. The case of chaperonins (proteins helping other proteins to fold inside cells) is not the general case and only shifts the problem (who folds the folders?). The possibility of sequentially denaturing/refolding a given protein clearly indicates that the three-dimensional structure is in some way encoded in its amino acid order [4,5].

The sequence-structure translation (or folding process) of the proteins can then be considered as a self-organizing process driven by the interaction of the polymer with the solvent. In other words, the folding process is basically a "solvation business" and the protein sequence can be considered as a recipe that the chemico-physical environment the protein is embedded into is able to decode and interpret [6]. As a matter of fact, to make a large molecule, as a protein is, soluble in water is not a simple affair: there are many different aspects to be taken into consideration like the delicate balance between the solvent loss of entropy due to the exposition of hydrophobic residues (this is the basis of the so called "hydrophobic bond" which is the energetically main driving force of protein folding), the amount of perturbation of the hydrogen bonding network encompassing both the solvent and solute, the local and non-local electrostatics effects like salt bridges and charge-charge interactions [2,7,8,9] . These intricacies explain why, notwithstanding the significant efforts by many scientists during these last fifty years, the sequence/structure prediction puzzle is still far to be definitively solved.

Obviously these efforts have not been completely useless and we have learned much about protein chemistry and physiology and, probably still more importantly, we have recognized that many apparently heterogeneous phenomena going from protein-protein aggregation and interaction, to allosteric regulations induced by specific effector drugs and hormones or by chemico-physical micro-environmental changes like pH modifications can be traced back to the "folding problem"[1].

Roughly speaking, the two main routes along which protein folding research has developed include:

a) Trying to crack the protein sequence-structure code by generating realistic models of how the principal protein folding driving forces do interact (*ab initio* route) [10].
b) Trying to derive empirical rules that could act as proxies of the sequence-structure code by looking for statistical regularities relating sequence and structure information in big data bases (*bioinformatics* route) [11].

There is a voluminous literature dealing with theoretical models following *ab initio* approaches to the sequence-structure puzzle [12]. The great majority of the theoretical models adopt a statistical physics perspective based on proteins considered as lattices, i.e., squared grids in which each residue is considered as interacting with the same number of neighbours [2.12]. These interactions are modeled by carefully chosen potential functions, generally based on hydrophobicity or related properties [12,13] associating an energetic score to each amino acid pair interaction. Dill and co-workers [14] suggested a simplified lattice model and used it to obtain minimal modeling of polypeptide chains. To account for the major interactions in proteins, the latter authors argued that the naturally occurring amino acids can be divided into hydrophobic and polar categories. Chan and co-workers emphasize that their model can capture many important features of proteins, including cooperativity, folding kinetics, and structural properties [15].

The bioinformatics philosophy adopts an alternative view of the sequence-structure puzzle: instead of looking for "universals", i.e., general laws linking sequence and structure across all protein families, they apply a purely local statistical approach [12]. A protein whose primary structure only is known is compared, in terms of relative sequence alignment, with a large number of proteins whose three-dimensional structure is solved. Scoring a significant (>30%) superposition between the query sequence and those of already solved proteins allows for structural inferences. This line of research is followed by many scientists with the impetus of post-genomic projects, whose basic aim is obtaining from genetic information structural models of the encoded proteins. A large number of research groups are presently investigating new algorithms to maximize the ability in discovering even relatively "remote" homologues of a given leader sequence.

The present contribution redounds to a "third route" located approximately half way between the two main ones [6]. The proposed approach is based on the consideration of protein sequences as numerical series obtained by the substitution of the residues with a carefully chosen interaction potential (typically hydrophobicity). These numerical series are investigated in terms of their internal periodicities that are considered as proxies of meaningful patterns for sequence/structure translation. Each protein is then described by a set of numerical invariants describing the amount (and shape) of hydrophobicity patterning along the primary structure. Different protein data sets are investigated for the ability of such a description to predict global properties of the protein molecules such as stability, folding mechanism, effects of mutations.

The consideration of proteins as numerical time series has a long history, dating back to the pioneering work by Zimmermann et al. [16] and Kyte and Doolitle [17]. These initial studies, while providing useful insights, were limited by the use of signal analysis methods not completely suitable for protein sequences. In fact Fourier analysis and linear autocorrelation functions (the basic methods used) have limitations for protein sequence studies, since they assume sequence stationarity and signals with a long sequence length typically longer than an average protein. Thus, although strictly periodic features can be identified; complicated, less obvious features are easily missed. Other methods attempted include the demonstration of a correlation between hydrophobicity patterning of peptides and their relative receptors by Mandell et al. [18] using singular value decomposition, and the use of wavelets to identify change points in trans membrane proteins.

This approach takes from the *ab initio* route the idea that the key of sequence/structure code must be looked for in the solvent/solute interaction by means of some sort of potential, while at the same time following the bioinformatics idea in looking at statistical regularities inside carefully chosen protein data sets.

## Recurrence Quantification Analysis

In this section we concentrate on the application of a method originally developed as a quantitative tool by Webber and Zbilut [19] following an initial qualitative proposal by Eckmann and Ruelle [20] for the analysis of time series. This method called Recurrence Quantification Analysis (RQA) was originally applied to protein sequences by our group allowing us to produce relevant results as for many aspects of protein science [6,21-23].

RQA by no means can be considered the only possible approach to time-series-like analysis of protein sequences. As has been mentioned, other approaches have given promising results in the same field like wavelets [6] or Singular Value Decomposition (SVD) [6]. Nevertheless the versatility and relative easiness of the RQA mathematical bases makes it a good primer for an introduction to the signal analysis perspective of protein science.

## How Proteins Appear from a Signal Analysis Perspective

When coded as mono-dimensional arrays of hydrophobicity values corresponding to the amino acid sequence, the primary structure of a protein can be considered as a numerical discrete series equivalent to a time series with the amino acid order playing the role of subsequent time intervals (Fig. 1).

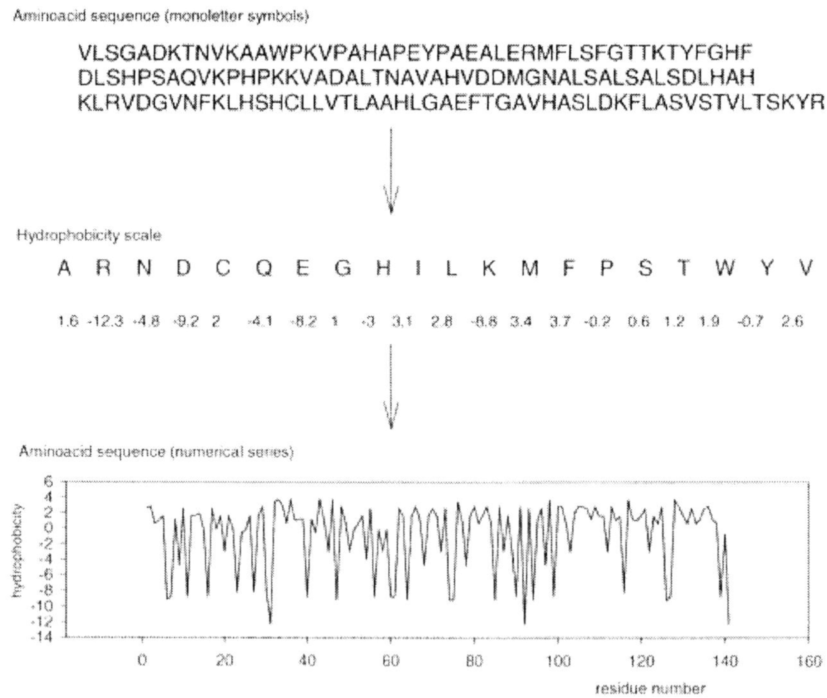


Figure 1. Transformation of protein letter code to hydrophobicity code.

Thus, on a purely formal point of view, any one of the myriads of techniques routinely used for signal analysis in electronics as well as in physiology or meteorology could be profitably applied to protein hydrophobicity sequences to extract potentially relevant features of primary structure. From a practical viewpoint, the fact that protein sequences are short with respect to the signals analysed in other fields (even if there is no reasonable physical limit to the length of a polypeptide chain, the greater part of naturally occurring proteins have less than 1000 residues) and in some cases extremely short (i.e. less than 50 residues) drastically limits the range of signal analysis techniques usable in this context. Moreover, protein hydrophobicity profiles are basically non-stationary signals displaying different statistical and correlation features along the chain, which does not favour classical techniques such as Fourier analysis.

## Methods

The ideal method for approaching signal analysis of protein sequences should be nonlinear, independent of any stationary assumptions, and able to deal with very short series [19,20]. Satisfying these constraints are those approaching the analysed series from a purely correlative point of view, with no *a priori* distributional and/or physical assumption. The only aim of these methods is to look for autocorrelation patterns along the series, i.e., for the recurrences of particular short motifs along the chain (as in recurrence quantification analysis, RQA) or for periodicities of no predefined functional form spanning all the studied sequences (as in singular value decomposition, SVD).At the basis of all these methods is the transformation of the original series into its "embedding matrix" with the method of delays [24].

The embedding procedure consists of building an $n$-column matrix (in the example below $n = 4$) out of the original linear array by shifting the series by a fixed lag. For example, given the series 10, 11, 21, 32, 41, 35, 40, 19..., the corresponding 4-dimensional embedding space at lag = 1 (the discrete character of amino acid sequences dictates this choice) is

10 11 21 32
11 21 32 41
21 32 41 35
32 41 35 40
etc., ...

The rows of the embedding matrix (EM) correspond to subsequent windows of length 4 (embedding dimension) along the sequence [6,19, 20,24]. Notice that the last $n$ values are eliminated from the analysis as an obvious consequence of shifting the series for the embedding. RQA is based on the computation of the Euclidean distance matrix (DM) between the rows (epochs) of the EM [24] looking for epochs close to each other (recurrences). SVD is based on the correlation matrix among EM columns, whereas wavelets are based on the correlation of the rows of the EM with user-defined kernels [6,24].

The choice of the embedding dimension corresponds to the choice of the scale at which the autocorrelation structure of hydrophobicity pattern is estimated, which varies across different techniques and problems. All the signal analysis techniques used in research on proteins give a global picture of the series in terms of degree of complexity (relative order/disorder of hydrophobicity distribution along the series), presence of singularities (regions within the sequence strongly different in terms of hydrophobicity pattern), and specific periodicities. In general, they quantitatively describe the shape of the hydrophobicity profiles by appropriate numerical indicators

Recurrence quantification analysis is a relatively new nonlinear technique, originally developed by Eckmann et al. [20] as a purely graphical method and then made quantitative by Webber and Zbilut [19]. It was successfully applied to different fields ranging from physiology [19,25] to molecular dynamics [26] and the study of chemical reactions [27]. Only in relatively recent times RQA was investigated by our group for its ability to deal with protein sequences [6].

The concept of recurrence is straightforward: for any ordered series (time or spatial), a recurrence is simply a point which repeats itself. In this respect, the statistical literature points out that recurrences are the most basic of relations [19] shaping a given system, since they are strictly local and independent of any mathematical assumption regarding the system itself. Furthermore, it is worth stressing that calculation of recurrences, unlike other methods such as Fourier, Wigner-Ville, or wavelets, requires no transformation of the data and can be used for both linear and nonlinear systems.

The concept of a recurrence can be expressed as follows: given a reference point, $X_0$, and a ball of radius $r$, a point $X$ is said to recur (with reference to $X_0$) if

$$B_r(X_0) = (X : \|X - X_0\|) \leq r \tag{1}$$

In the case of a time series, i.e., of a system occupying in different times different positions along a trajectory in a suitable state space, the recurrences correspond to the time points where the system passes nearby to already visited states. In the case of protein sequences, time corresponds to the amino acid order and the recurrences are patches, with a length equal to the embedding dimension, sharing their hydrophobicity profile with other patches along the chain. The number and relative positions of recurrences are expressed by recurrence plots (RP) that are symmetrical $N \times N$ arrays in which a point is placed at $(i, j)$ whenever a point $X_i$ on the trajectory is close to another point $X_j$. The closeness between $X_i$ and $X_j$ is expressed by calculating the Euclidian distance between these two normed vectors, i.e., by subtracting one from the other obtaining the expression $\|X_i - X_j\| \leq r$ where $r$ is a fixed radius. If the distance falls within this radius, the two vectors are considered to be recurrent, and graphically this can be indicated by a dot (Figure 2).

Thus, recurrence plots simply correspond to the distance matrix between the different epochs (rows of the embedding matrix) filtered, by the action of the radius, to a binary 0/1 matrix having a 1 (dot) for distances falling below the radius and a 0 for distances greater than radius. Distance matrixes are demonstrated to convey all the relevant information for the global reconstruction of a given system [28]. An important feature of such matrixes is the existence of short line segments parallel to the main diagonal, which correspond to sequences $(i, j)$, $(i + 1, j + 1)$, ..., $(i + k, j + k)$ such that the fragment $X(j)$, $X(j + 1)$, $X(j + k)$ is close to $X(i)$, $X(i + 1)$, ..., $X(i + k)$. The absence of such patterns suggests randomness. For protein sequences these deterministic lines correspond to contiguous patches of similar hydrophobic/hydrophilic patterns.

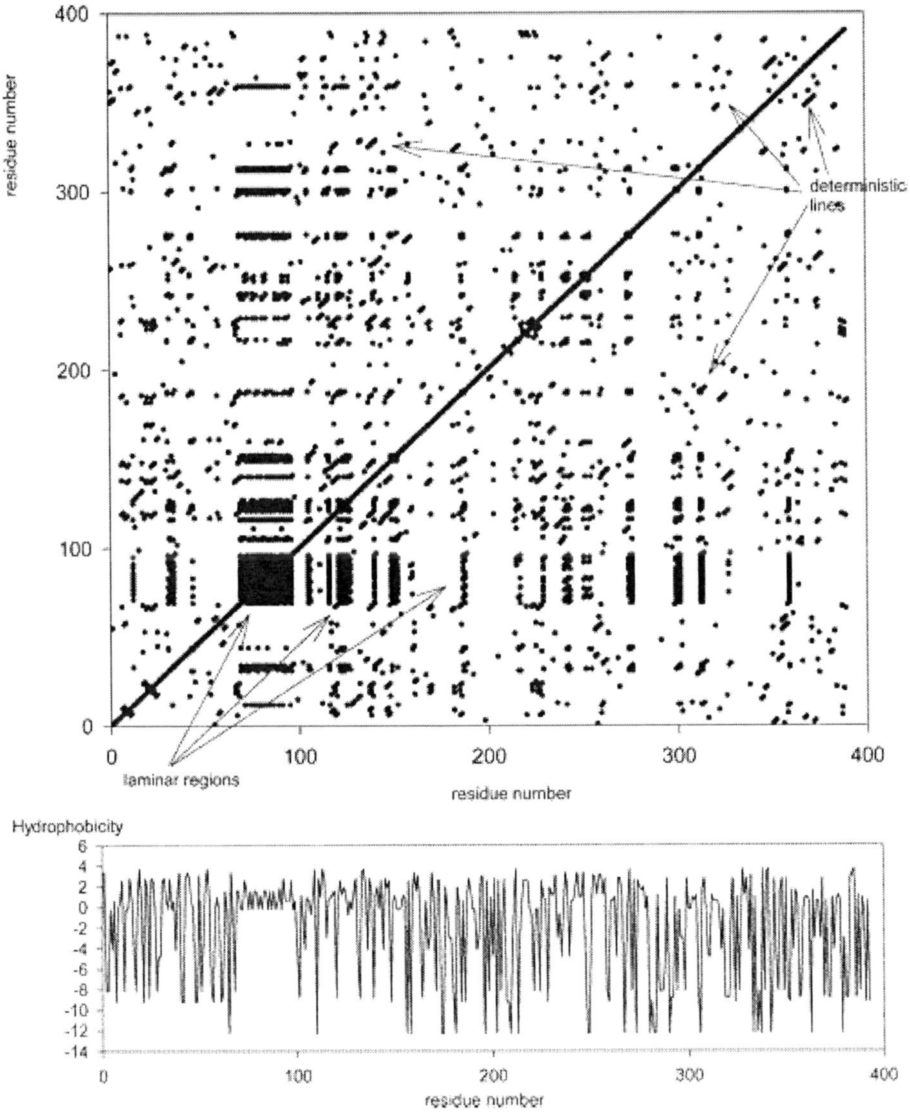


Figure 2. Hydrophobicity coded protein (bottom), and its recurrence plot with important features labeled.

Because graphical representations may be difficult to evaluate, Zbilut and Webber [19] developed several strategies to quantify features of such plots originally pointed out by Eckmann et al. [20]. The closer inspection of the RPs reveals small-scale structures (the texture), which are *single dots*, *diagonal lines* as well as *vertical* and *horizontal lines* (the combination of vertical and horizontal lines obviously forms rectangular clusters of recurrence points). These structures are taken into account and quantified by the descriptors used in RQA. In particular:

- *Single, isolated recurrence points* can occur if states are rare, if they do not persist or fluctuate heavily. However, they are not a unique sign of chance or noise. These dots are counted in the quantitative descriptor recurrence (REC).

• A *diagonal line* $\mathbf{R}_{i+k,j+k} = 1$ (for $k = 1...l$, where $l$ is the length of the diagonal line) occurs when a segment of the numerical series runs parallel to another segment, i.e., the sequence visits the same region of the phase space at different intervals. The length of this diagonal line is determined by the duration of such similar local evolution of the segments. These diagonal structures are called deterministic lines and are counted by the descriptor determinism (DET), entropy (ENT) and maxline (MAXL).

• A *vertical (horizontal) line* $\mathbf{R}_{i,j+k} = 1$ (for $k = 1...v$, where $v$ is the length of the vertical line) marks a length in which a state does not change or changes very slowly. It seems, that the state is trapped. These are called laminar regions and are counted in the descriptors laminarity (LAMIN) and traptime (TRAPT).

Figure 2 represents a typical RP of a protein (human P53) with the indication of the different features taken into consideration by quantitative descriptors that in turn are defined in Table 1

**Table 1. RQA definitions**

| Measure | Definition |
|---|---|
| Recurrence , *REC* | Percentage of recurrence points in an $RR = \frac{1}{N^2} \sum_{i,j=1}^{N} R_{i,j}$ RP: |
| Determinism, *DET* | Percentage of recurrence points which form diagonal lines: $DET = \frac{\sum_{l=l_{\min}}^{N} lP(l)}{\sum_{i,j}^{N} R_{i,j}}$ *P(l)* is the histogram of the lengths $l$ of the diagonal lines. |
| Laminarity, *LAM* | Percentage of recurrence points which form vertical lines: $LAM = \frac{\sum_{\upsilon=\upsilon_{\min}}^{N} \upsilon P(\upsilon)}{\sum_{\upsilon=1}^{N} \upsilon P(\upsilon)}$ *P(v)* is the histogram of the lengths $v$ of the vertical lines. |
| Entropy, *ENT* | Shannon entropy of the distribution of the diagonal line $ENT = -\sum_{l=l_{\min}}^{N} p(l) \ln p(l)$ lengths *p(l)*: |
| Trend, *TREND* | Paling of the RP towards its edges: $TREND = \frac{\sum_{i=1}^{N-2}[i-(N-2)](RR_i - \langle RR_i \rangle)}{\sum_{i=1}^{N-2}[i=(N-2)/2]^2}$ |
| Trapping time, *TRAPT* | $TT = \frac{\sum_{\upsilon=\upsilon_{\min}}^{N} \upsilon P(\upsilon)}{\sum_{\upsilon=\upsilon_{\min}}^{N} P(\upsilon)}$ Average length of vertical lines: |
| Longest diagonal line, MAXL | Length of the longest diagonal line: MAXL $L_{\max} = \max(\{l_i; i = 1 \ldots N_l\})$ |

At the end of the day a protein will be described by a 9 components vector corresponding to the 7 RQA (order dependent) descriptors reported in Table 1 plus two statistical (order independent) descriptors corresponding to the mean and standard deviation respectively of the hydrophobicity series [21].

A collection of proteins will then correspond to a unit/variable matrix in which the proteins are the rows (statistical units) and the columns (variables) the dynamical and statistical descriptors representing different aspects of the hydrophobicity patterning along the chain. From this point on, the analysis is not different from any other statistical study, so we can try and discriminate different classes of objects (let's say mesophylic/thermophylic proteins, two-state/multi-state folders) on the basis of their differences in descriptor values or we can adopt an unsupervised attitude and simply look among variable correlations to find unexpected regularities between different aspects of hydrophobicity patterning description.

In the following we will analyse some applications of RQA to protein studies trying to highlight the mixed descriptive/mechanistic character of the approach.

# Case Studies

## 1. Are there Some General Syntactic Invariants in Protein Hydrophobicity Patterning ?

Protein sequences are, with rare exceptions (e.g. fibrous polymerizing proteins such as collagen or silk), quasi-random strings of symbols with scant evidence of order or periodicity: a reliable estimate of the entropy reduction due to the autocorrelation of residues in an average protein sequence is only about 1% [29]. Nevertheless, such quasi-random strings are the basic recipes producing refined three-dimensional structures, which sustain sophisticated dynamics along with specific physiological roles. Thus, the observed quasi-randomness may be a specious image for underlying meaning. It is interesting to note that a similar situation occurs in the case of human languages where it is almost impossible to generate meaningful texts using just periodic repetitions of symbols [30]. There is, however, a fundamental difference between linguistic rules and the rules governing sequence/structure/activity of proteins: in human languages the linkage between the strings of characters (words) and their semantic meaning is completely arbitrary and needs an external intelligent and active receiver to be decoded. Amino acid sequences, on the other hand, are translated into biologically meaningful messages in the form of proteins by the physicochemical environment (e.g., ionic strength, relative hydrophobicity, temperature, pressure). At the same time, the quasi-random qualification of symbolically coded protein sequences evokes the possibility of solving the sequence/structure/activity puzzle by discovering subtle, albeit crucial regularities in the juxtaposition of symbols.

The importance of such regularities in decoding signals can be better appreciated if one recalls the development of cryptography during the Second World War. The decipherment of hidden information in encrypted messages was based upon the notion that any human language, despite its apparent randomness and arbitrariness, is endowed with regularities of various kinds and that no "masking code" can obscure the code-independent features typical of the original language [30].

In other words, at the basis of each code-cracking strategy is the notion of the existence of a unique meaningful original message conveying a specific semantic information (e.g. "the ship will leave at 8 AM from the pier 42 of Malta harbour") that is made obscure by masking codes. What remains alive (and invariant) of the original message across the different coding strategies? The answer is: a probabilistic distribution of word lengths (short words will be more frequent than long ones irrespective of the used code), letter positional correlations (each language has some favoured and disfavoured neighbouring letters) and other subtle deviations from symmetry that can help to unveil the original message. Thus we looked for the presence of such among-codes invariants across different coding strategies of the same set of proteins. Basically each sequence of a 1141 proteins data set was transformed into 7 distinct numerical series plus a symbolic series corresponding to the one-letter code of residues. The adopted codes corresponded to the following physicochemical properties of amino acids: three hydrophobicity scales (Chothia [31], Kyte and Doolittle [17], Miyazawa-Jernigan [32]), molecular weight, volume [16] polarity [33] and molar refractivity [34].

For each 1141 x 8 series corresponding to the proteins coded by physico-chemical descriptors plus the symbolic code, the RQA REC and DET parameters were computed (the symbolic code too allows for RQA formalization given we use a radius equal to zero for the impossibility of any continuous distance measure) and the principal component analysis (PCA) was applied to both REC and DET formalizations to elucidate the between codes correlation structure.

When submitting the RQA-based representations of proteins of the various codings to a Principal Component Analysis (PCA) [6], we should obtain, if our hypothesis of a basic code-independent structure is true, as main mode (first Principal Component, PC1), a consensus axis collecting all the codes whose eigenvalue represents the degree of code-independent autocorrelation structure; namely, the image of the code-independent meaning conveyed by the message. This was actually the case: all the codes were strongly loaded on the first principal component which, both for REC and DET, was the most important source of information explaining, respectively, 70% and 50% of the total variability (Table 2). It is worth noting that, the symbolic coding was highly correlated with the first component, as a further indication of the role of code-independent autocorrelation measure played by PC1.

Recalling the fact principal components correspond to the mutually independent correlation structures present in the data, we can immediately recognize PC1 as the image of the "code invariant" structure generated by the above mentioned syntactic regularities. In fact all the different codes are positively loaded on PC1 with the symbolic code (by definition neutral to any chemico-physical definition) being heavily loaded on PC1. The eigenvalue of PC1 is by far the most important one, corresponding to 69% and 50% of total information for REC and DET respectively, pointing to the code-independent amount of order in amino acid distribution along the chain as the main order parameter shaping protein sequences variability.

When the proteins are shuffled by random mixing of residue positions along the chain, we obtain completely different PC1DET ordering of the protein data set, indicating the patterns of hydrophobicity distribution have something to do with the protein individuality despite their basically non-periodic character, i.e. the juxtaposition of residues is not simply a matter of chance. Thus we are now interested in ordering the different sequences in terms of PC1DET (amount of order in the hydrophobicity pattern) to discover a possible link between hydrophobic patterning along the chain and protein physiology.

**Table 2. Loadings of the Amino Acid Codes on the First Three Principal Components from RQA Filtered Proteins[a]**

| (a) PCA on Recurrence | | | |
|---|---|---|---|
| **code** | **PC1REC** | **PC2REC** | **PC3REC** |
| Chothia Hydrophobicity | 0.86 | -0.20 | -0.07 |
| Kite – Doolittle Hydrophobicity | 0.82 | 0.04 | 0.42 |
| Miyazawa – Jernigan Hydrophobicity | 0.77 | -0.29 | 0.37 |
| Molecular weight | 0.90 | -0.23 | -0.26 |
| Polarizability | 0.69 | 0.60 | -0.20 |
| Molar Refractivity | 0.91 | -0.03 | -0.17 |
| Volume | 0.79 | 0.39 | 0.21 |
| Symbolic coding | 0.92 | -0.14 | -0.25 |
| % expl.variance | 69.9 | 8.9 | 6.9 |
| **(b) PCA on Determinism** | | | |
| **code** | **PC1DET** | **PC2DET** | **PC3DET** |
| Chothia Hydrophobicity | 0.71 | -0.04 | 0.38 |
| Kite-Doolittle Hydrophobicity | 0.77 | -0.30 | -0.19 |
| Miyazawa – Jernigan Hydrophobicity | 0.64 | -0.42 | 0.42 |
| Molecular weight | 0.68 | 0.44 | -0.01 |
| Polarizability | 0.79 | -0.19 | -0.15 |
| Molar Refractivity | 0.65 | 0.49 | 0.26 |
| Volume | 0.70 | 0.001 | -0.50 |
| Symbolic coding | 0.70 | 0.10 | -0.12 |
| % expl.variance | 50 | 9.4 | 8.9 |

Panels (a) and (b) refer, respectively, to REC and DET variables and contain the "loadings" (correlation values) of the original variables with the new ones extracted by the PCA algorithm. The Principal Components were obtained from matrices containing as rows the 1141 proteins of our data set, and as columns the REC (panel a) and DET (panel b) descriptors of each protein calculated from the profiles in the various amino acid codings. The last rows in both panels contain the % of total variance explained by each component.

When ordered in terms of increasing PC1DET (the global, code-independent hydrophobicity regularity measure) we obtained a highly skewed distribution with the by far great majority of sequences lying in the PC1DET=-2, PC1DET=+2 interval (remember the component has zero mean and unit standard deviation by construction) plus a 50 elements "tail" of extremely ordered proteins reported in Table 3.

**Table 3. Elements of the "High Determinism Tail" in the Distribution along the First Determinism Component (PC1DET) of the Protein Data Set Used in This Work**

| Swiss-Prot code | Name | PC1DET |
|---|---|---|
| P35324 | CORNIFIN ALPHA | 8.52 |
| Q62267 | CORNIFIN B | 7.79 |
| Q63532 | CORNIFIN ALPHA | 6.95 |
| Q62266 | CORNIFIN A | 6.19 |
| Q07187 | EM-like PROTEIN GEA1 | 6.05 |
| P06144 | LATE HISTON H1 | 5.27 |
| P35326 | SMALL PROLINE-RICH PR. 2A | 4.97 |
| P17483 | HOMEOBOX PROTEIN HOX-B4 | 4.49 |
| 035762 | HOMEOBOX PROTEIN NKX-6.1 | 4.32 |
| P37108 | SIGNAL RECOGN. PART. 14 Kda | 4.24 |
| P28318 | PROTEIN MRP-126 | 4.19 |
| P15771 | NUCLEOLIN | 3.99 |
| 009116 | CORNIFIN BETA | 3.96 |
| P02604 | MYOSIN LIGHT CHAIN 1 | 3.89 |
| P42132 | SPERM PROTAMINE P1 | 3.79 |
| P22793 | TRICHOHYALIN | 3.71 |
| Q34522 | NADH-UBIQ. OXYDORED. CHAIN 3 | 3.61 |
| P17502 | PROTAMINE | 3.52 |
| P42129 | SPERM PROTAMINE P1 | 3.42 |
| *P*22238 | DESICCATION REL. PROT. | 3.37 |
| Q22053 | FIBRILLARIN | 3.35 |
| P55947 | COPPER-METALLOTHIONEIN | 3.30 |
| P15870 | HISTONE H1-DELTA | 3.21 |
| P41139 | DNA BINDING PROT. INHIB. ID-4 | 3.13 |
| Q13329 | COMPLEXIN 2 | 3.08 |
| Q63754 | BETA-SYNUCLEIN | 3.07 |
| Q01821 | GUANINE NUCL. BIND. | 3.07 |
| P34618 | CEC-1 PROTEIN | 3.04 |
| P06146 | HISTONE H2B.2, SPERM | 3.02 |
| P09442 | LATE EMBRYOG. PROT. D-11 | 3.01 |
| P02292 | HISTONE H2B.3, SPERM | 2.99 |
| P12950 | DEHYDRIN DHN1 | 2.97 |
| Q05831 | SPERM-SPECIFIC PROTEIN PHI-2B | 2.79 |
| P12952 | DEHYDRIN DHN2 | 2.77 |

**Table 3. continued**

| Swiss-Prot code | Name | PC1DET |
|---|---|---|
| P47928 | DNA BIND. PROTEIN INHIB. ID-4 | 2.74 |
| P52168 | GATA-BINDING FACTOR-A | 2.74 |
| P22974 | SPERM SPECIFIC PROTEIN PHI-2B | 2.66 |
| P12035 | KERATIN TYPE II CYTOSKEL. 3 | 2.62 |
| Q09821 | SPERMATID NUCLEAR TRANS. | 2.56 |
| Q15672 | TWIST RELATED PROTEIN | 2.46 |
| P90648 | MYOSIN HEAVY CHAIN KINASE B | 2.40 |
| P06145 | HISTONE H2B.1, SPERM | 2.32 |
| P02836 | SEGMENT. POLAR. HOMEOBOX | 2.31 |
| 042105 | COMPLEXIN 2 | 2.24 |
| P17480 | NUCLEOLAR TRANSCR. FACT. 1 | 2.23 |
| P54844 | TRANSCR. FACTOR MAF | 2.20 |
| P40262 | HISTON H1 E | 2.20 |
| P25979 | NUCLEOLAR TRANSCR. FACTOR 1 | 2.17 |
| P21952 | OCT. BIND. TRANSCR. FACT. 6 | 2.07 |
| Q12948 | FORK HEAD BOX PROTEIN C1 | 2.04 |

All the extremely deterministic proteins reported in Table 3 share the property of being involved in protein-protein or DNA-protein interactions, both for regulatory and structural purposes (e.g., histones, protamines, and trascription factors) as well as for forming polymeric assemblies (cornifin, myosin, keratin).

Recently Dunker and co-workers demonstrated how the most represented class of natively unfolded structures is composed of polypeptides involved in protein-protein interactions. Moreover, the increasing evidence that low complexity sequences tend to be natively unfolded [3,35] suggests a check for the presence of an excess of natively unfolded zones in the deterministic tail of our data set. The 10 most deterministic sequences scored a percentage of estimated disorder (computed by the PONDR predictor [3]) of 66.46% against the 27.27% of the 10 proteins situated at the low determinism tail (significance of $p < 0.001$). Calculation of a foldability coefficient [35] for the highly deterministic sequences listed in Table 3 indicates that more than 75% may be classified as natively unfolded, and this figure becomes even larger if the reduced form of the -S-S- bridges present in many sequences is considered. From the above analysis the role of "deterministic spots" as crucial sites for interaction seems to gain support. Assuming that protein-protein interactions are driven by essentially the same type of forces leading to mutual recognition between different portions of the same molecule in normal folding, we can hypothesize that highly deterministic sections along the sequence mark the nucleation zones for both folding and protein-protein interactions. Thus an albeit very crude but relevant aspect of protein sequence/structure code can be learned by this analysis: extremely repetitive sequences point to natively unfolded, interaction prone systems.

We have demonstrated the presence of code-independent features of primary structures syntaxys; however, the seven adopted codes are not fully equivalent as for their ability to detect repetitive motifs along protein chains. The amount of both recurrence and determinism relative to the Myjazawa-Jernigan hydrophobicity is significantly higher with respect to the other codes. This is probably due to the "nearness" of this coding scheme to the real (and still unknown) Rosetta stone linking sequence to structure. This made the Myjazawa-Jernigan hydrophobicity the preferred code for other investigations in the signal analysis approach to protein sequence/structure relations.

## 2. Can We Move from the Population to the Single Molecule Level?

In the above example we adopted a purely descriptive attitude: relying on a very big and diverse protein population, without any *a priori* bias, we let the data structure to emerge by the action of a simple correlation strategy. Discovering that the extreme values of the principal component corresponding to the amount of determinism of hydrophobicity patterning along the chain were largely natively unfolded, aggregation-prone proteins, allowed us to hypothesize a relation between the presence of ordered patches of hydrophobicity and aggregation propensities of proteins. Knowing that aggregation process is preceded by a partial unfolding of the portion of molecule undergoing aggregation suggests to us a deeper understanding of the dynamic/sequence relation: the most deterministic (or laminar) portion of a protein are the ones exhibiting a richer dynamical behaviour, i.e. are the most flexible portions of the molecule.

It is worth noting we are moving from a purely population-based perspective (the one generally adopted by bioinformatics approaches) to a more mechanistic one, shifting by statistical evidence from a large set of proteins to the actual behavior of a single protein. An almost ideal model system to confirm these findings is represented by the fragment 1-40 (AB40) of APP (amyloid precursor protein) thought to be implicated in Alzheimer's disease, whose aggregation propensity has been carefully analysed by many authors (for a review see [36]).

The molecular dynamics of AB40 was simulated at three pH levels [Low (AB40L), Medium (AB40M), High (AB40H)] corresponding to three experimentally demonstrated different aggregation propensities, with M being the more aggregating condition corresponding to the isoelectric point of AB40 where the zero net charge of the system does not allow the electrostatic repulsion to protect the protein from aggregation [21]. Molecular dynamics simulation confirmed the link between aggregation propensity and conformational flexibility showing the medium pH condition (AB40M) was endowed with an extremely large conformational flexibility with respect to the two other conditions [37]. We can appreciate the link between conformational flexibility and deterministic patches: as can be seen from Fig.3 which connects the molecular dynamics simulation (expressed as the RMSD of the single residues) and RP, laminar areas of the RP (roughly 5–15; 20–30) exhibit the largest number of conformational changes. This is to say the patchy areas are significant sources for the motions. The conjecture was confirmed by a Pearson correlation analysis of the three MD simulations with the laminar patches (AB40L, $r = 0.735$, $p = 0.001$; AB40M, $r = 0.555$, $p = 0.007$; AB40N, $r = 0.569$, and $p = 0.005$; Bonferroni adjusted probabilities, see also Fig.4).

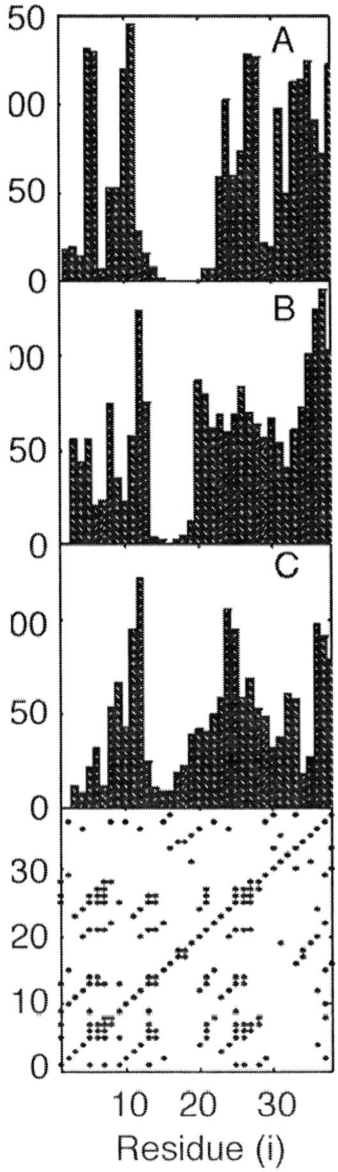


Figure 3. Recurrence plot (bottom) and histograms for the three simulated pH values of the RMSD values of the simulation. Note the corresponce between the laminar area of the RP (square patches) and high values of the histograms.

What are the consequences of the presence of short, singular patches in the hydrophobicity pattern of consecutive residues for the dynamics of the protein structure from a theoretical point of view? We have argued that there exist many physical and biological motions which may be better modeled by non-Lipschitz (nonsmooth) differential equations whereby there is no single solution to the equation. In plain terms, a nonsmooth (and thus non-Lipschitz) approach implies the existence of discrete and identifiable turning-points in which the observed phenomenon instead of approaching in an infinite time to a limit (as in classical differential equations), on the contrary, is abruptly stopped. Examples of these kinds of dynamics are the "rip" of a flag or the discharge of a seismic wave. In all these systems we can identify singular points in which the dynamics abruptly "forget the past" and re-start with a completely stochastic choice of direction [38]. In this situation the relative probability of moving one way or another (in terms of folding/aggregation) becomes a combinatorial one with a resultant *nondeterministic* dynamics, and a *stochastic attractor*. This is to say that there exist unstable singularities with an associated probability distribution regulated by factors unique to the given system [38]. In the protein case, narrow patches of high determinism and/or laminarity at low radius are the basis of this singularity [21, 22]. Although in the present case the discussion centers around a topological feature, without loss of generalization, it can be argued that the energy functions of molecular dynamics behave in a similar way. As a matter of fact hydrophobicity is a partition function governing the relative propensity of being solvated and thus the hydrophobicity profile can be considered as the static recipe giving rise to dynamical behavior of the protein in solution. This is to say that if the process is described as a minimization of the energy function through potential minima, the boundaries can be described as unstable singularities as well [38]. In the long run, the associated probabilities that may govern whether a given protein will go to its native fold or an aggregation, may depend upon its TREND/LAM characterization.

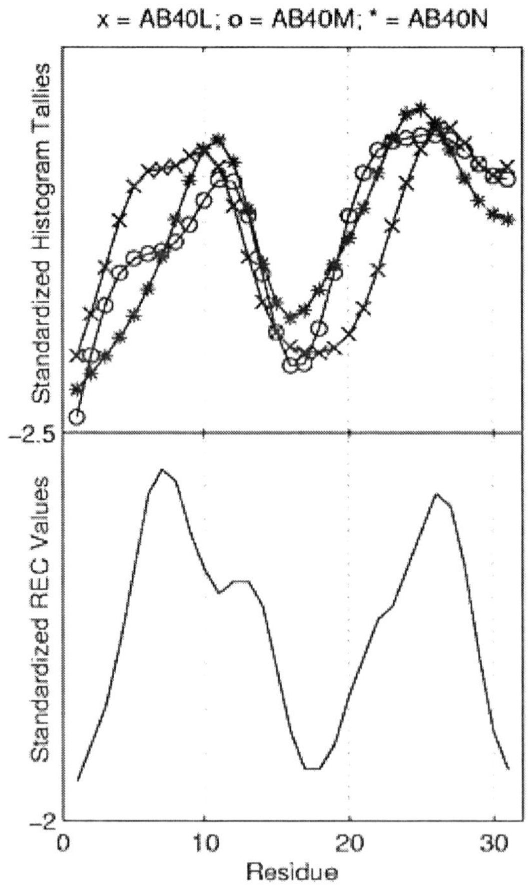


Figure 4. Standardized histogram values of Fig. 3 (top) and standardized REC values (bottom). See text.

The link between laminar and deterministic patches along the chain and aggregation phenotype can be fully appreciated in those structural proteins, like dragline silks, whose strong peculiarity resides in the formation of macroscopic structures endowed with astonishing features of elasticity and strength. At odds with classical globular proteins, the constituents of dragline silk like the ADF-3 protein whose RP is reported in Fig.5 exhibit a quasi periodic patterning of hydrophobicity with a clear modular architecture supporting the generation of the multimeric fibers constituting draglines.

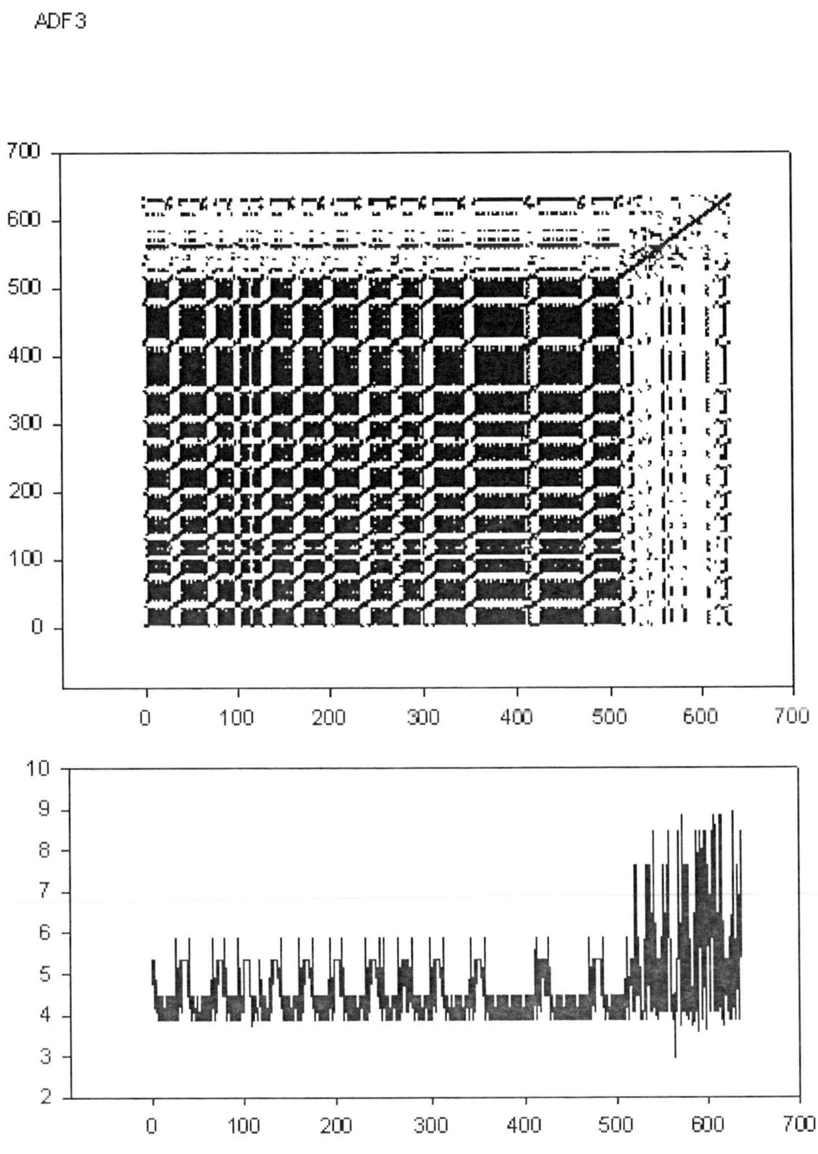


Figure 5. Recurrence plot (top) and hydrophobicity of laminar areas of ADF-3. Note the quasi-periodic laminar areas followed by an irregular "non-repetitive" region.

What distinguishes silk proteins from many other protein sequences is the marked periodicity of the residues along the sequence. This property is shared by numerous proteins involved in the construction of large matrices such as collagen or keratin, and under this aspect makes these molecules more similar to artificial polymers than classical globular proteins. Prior to their exiting from the spider, they are highly soluble, which implies a strong "anti aggregation" feature embedded in silk protein sequences. Studies investigating different synthetic spider silk proteins have revealed that solubility is mainly determined by their repetitive regions which represent the largest part of dragline silk proteins [39]. It could also be shown, that factors, such as acidification and increase of phosphate concentration, which promote silk assembly *in vivo,* generally decrease solubility of synthetic silk proteins *in vitro*, an effect that is enhanced by a non-repetitive (NR) amino acid sequence located at the carboxyl terminus of dragline silk proteins. This indicates that the NR region might play an important role in the assembly of spider silk proteins.

Consequently, it follows that in order to understand the peculiar behavior of silks, the quantitative role of the amount of periodicity needs to be established. In line with the previous line of reasoning we investigated ADF-3 and its related ADF-4. Interestingly, what evolved was the demonstration that for folding and/or solubility relative to the 2 proteins to be effected, the NR regions appeared to be necessary through a type of "stochastic resonance." This is to say that the NR regions were effectively stochastic (lacking in any correlated structure), which feature "amplified" the laminar areas of the repetitive regions along with their hydrophobicity. It is hypothesiszed that this amplification provides just sufficient extra hydrophobicity to effect the change in solubility [40].

Following an analogous line of reasoning we can try and sketch a sort of empirical "kernel" for predicting the effect of a single mutation relative to the aggregation propensity of a given protein system [41]. The above described correspondence between highly flexible (and consequently potentially aggregation prone) and laminar/deterministic patches of AB40 peptide was complemented by the analysis of Acylphospatase (AcP) mutational spectra as for the modification of aggregation propensity. A significant finding included a correspondence between short deterministic patches of hydrophobicity distribution along the amino-acid sequence, what we term laminarity, [LAM, (*L*)], with 3D "unstructured" portions of acylphosphatase (AcP) . It was shown that the "ruggedness" of the hydrophobicity as measured by the derivative of hydrophobic change [what we term TREND, (*T*)], coincided with Dunker's "disorder" index [3]. The implication is that these areas may be inherently unstable, and somehow involved in the promotion of (at least) partial unfolding and aggregation. The degree to which these conditions exist probabilistically determines the propensity for aggregation. What was not determined is the effect of total charge on the probabilities. Thus we developed a kernel on the basis of different experimental evidences linking RQA descriptors and charge information. The kernel has the form:

$$Agg(rate) = Const. + a(|T|^{L^{TT}} / RR) * (Q / Length^2) \tag{2}$$

where *Const.* is a constant, a is an adjustable parameter, *T* s the TREND, *L* is the LAMINARITY, *TT* the TRAPPING TIME, *RR* the RECURRENCE RATE, and *Q* the net molecular charge. The data sets of acylphosphatase with their aggregation rates were compared to the formula, and were found to be significantly related to the function ($r - 0.79$; $p = 0.0002$) (Fig.6).

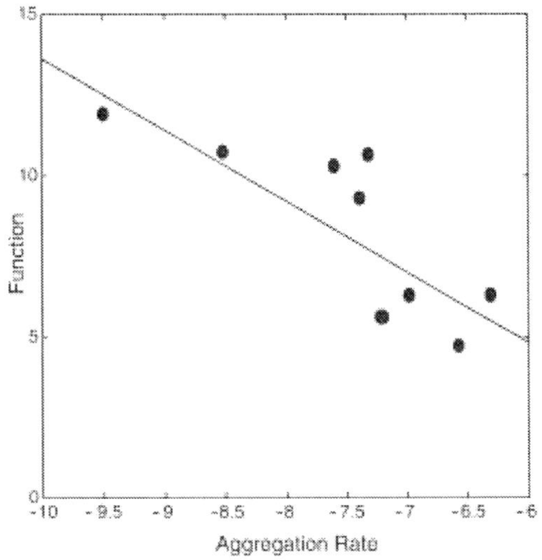


Figure 6. Plot of the ansatz of Eq. 2 vs. observed aggregation rates of AcP mutational aggregation rates.

Fig.6 has on the abscissa the observed aggregation rate of different single residue mutants (the points in the graph) and the ordinate displays Eq. 2. The empirical concordance of experimental data and theoretical equation opens the way to the possibility to sketching quantitative laws (at least variational) for aggregation and folding prediction.

## Conclusions

The above sketched case studies have in common the idea of deriving a synthetic description of different protein molecules based on the specific patterning of hydrophobicity distribution along the chain. These studies are based on a stochastic assumption; i.e., that protein folding occurs in the context of a probabalistic dynamic. To speak of "laws" probably makes little sense since the implied determinism cannot be easily demonstrated. Despite their roughness these descriptors have been shown to provide meaningful results in many different circumstances. Going more in detail, the basic idea at the basis of RQA protein sequence analysis, is to be able to detect (and count) all the autosimilarities inside the protein chain thus to be able to create a sequence alignment strategy based upon a physically associated metric.

## References

[1] Dobson C (2003). Protein folding and Disease: a view from first Horizon Symposium. *Nature* **2**: 154-160.

[2] Finkelstein AV, Galzitskaya OV (2004). Physics of protein folding. *Physics of Life Rev.* **1**: 23-56.

[3] Dunker KA, Brown CJ, Lawson DJ, Iakoucheva LM, Obradovic Z (2002). Intrinsic disorder and protein function. *Biochemistry* **41**: 6573-6582.

[4] Branden CL, Tooze J (1991). *Introduction to Protein Structure*. New York: Garland.

[5] Anfinsen CB (1973). Principles that govern the folding of protein chains. *Science* **181**: 223-230.

[6] Giuliani A, Benigni R, Zbilut JP, Webber CL, Sirabella P, Colosimo A (2002). Nonlinear signal analysis methods in the elucidation of proteins sequence-structure relationships. *Chem. Revs.* **102**: 1471-1491.

[7] Fernandez A, Scott R (2003). Dehydron: a structurally encoded signal for protein interaction. *Biophys. J.* **85**: 1914-1928.

[8] Lund M, Jonsson B (2005). On the charge regulation of proteins. *Biochemistry* **44**: 5722-5727.

[9] Miao J, Klein-Seetharaman, Meirovitch H (2004). The optimal fraction of hydrophobic residues required to ensure protein collapse. *J. Mol. Biol.* **344**: 797-811.

[10] Klepeis JL, Wei Y, Hecht MH, Floudas CA (2005). Ab initio prediction of three-dimensional structure of a de novo designed protein: a double-blind case study. *Proteins* **3**: 560-570.

[11] Cozzetto D, Tramontano A (2005). Relationship between multiple sequence alignment and quality of protein comparative models. *Proteins* **1**: 151-157.

[12] Tramontano A (2004). Integral and differential form of the protein folding problem. *Physics of Life Rev.* **1**: 103-127.

[13] Kuwajima K, Arai M (1999). *Old and New Views of Protein Folding*. Amsterdam: Elsevier.

[14] Dill KA, Bromberg S, Yue K, Fiebig KM, Yee DP, Thomas PD, Chan HS (1995). Principles of protein folding- a perspective from simple exact models. *Protein Sci.* **4**: 561-602.

[15] Chan HS, Bromberg S, Dill KA (1995). Models of cooperativity in protein folding. *Philos. Trans. R. Soc. Lond. B Biol. Sci.* **29**: 61-70.

[16] Zimmerman JM, Eliezer N, Simha R (1968). The characterization of amino acid sequences in proteins by statistical methods. *J. Theor. Biol.* **21**:170-201.

[17] Kyte J, Doolitle RF (1982). A simple method for displaying the hydropathic character of a protein. *J. Mol. Biol.* **157**, 105-132.

[18] Mandell AJ, Selz KA, Shlesinger MF (2000). Protein binding predictions from amino acid primary sequence hydrophobicity. *J. Mol. Liquids.* **86**:163-171.

[19] Webber CL, Zbilut JP (1994). Dynamical assessment of physiological systems and states using recurrence plot strategies. *J. Appl. Physiol.* **2**: 965-973.

[20] Eckmann JO, Kamphorst O, Ruelle D (1987). Recurrence plots of dynamical systems. *Europhys. Lett.* **4:** 973-975.

[21] Zbilut JP, Giuliani A, Colosimo A, Mitchell JC, Colafranceschi M, Marwan N, Webber CL, Uversky V (2004). Charge and hydrophobicity patterning along the sequence predicts the folding mechanism and aggregation of proteins: a computational approach. *J. of Proteome Res.* **3**: 1243-1253.

[22] Zbilut JP, Colosimo A, Conti F, Colafranceschi M, Manetti C, Valerio MC, Webber Jr CL, Giuliani A (2003). Protein aggregation/folding: the role of deterministic singularities of sequence hydrophobicity as determined by nonlinear signal analysis of acylphosphatase and Aβ(1-40). *Biophys. J.* **85**: 3544-3557.

[23] Porrello A, Soddu S, Zbilut JP, Crescenzi M, Giuliani A (2004). Discrimination of single aminoacid mutations of the P53 protein by means of recurrence quantification analysis. *Proteins* **55**: 743-755.

[24] Broomhead DS, King GP (1987). Extracting qualitative dynamics from experimental data. *Physica D* **20**: 211-226.

[25] Giuliani A, Piccirillo G, Marigliano V, Colosimo A (1998). A Non-linear explanation for aging induced changes in heartbeat dynamics. *Am. Journ. Physiol.* **275**: H1455-H1461.

[26] Manetti C, Ceruso MA, Giuliani A, Webber CL, Zbilut JP (1999). Recurrence quantification analysis as a tool for the characterization of molecular dynamics simulations. *Phys. Rev. E* **59**: 992-998.

[27] Masia M, Bastianoni M, Rustici M (2001). Recurrence quantification analysis of spatio-temporal chaotic transient in a closet unstirred Belousov-Zhabotinski reaction. *Phys. Chem. Chem. Phys*. **3**: 5516-5520.

[28] Thiel M, Romano M, Kurths J (2004). How much information is contained in a recurrence plot? *Phys. Lett. A* : 343-349.

[29] Weiss O, Jimenez-Montano M, Herzel HP (2000). Information content of protein sequences. *J. Theor. Biol.* **206**: 379-386.

[30] Colafranceschi M, Colosimo A, Zbilut JP, Uversky VN, Giuliani A (2005). Structure-related statistical singularities along protein sequences: a correlation study. *J. of Chemical Information and Modeling* **45**: 183-189.

[31] Chothia C (1976). The nature of the accessible and buried surfaces in proteins. *J. Mol. Biol.* **105**: 1-14.

[32] Miyazawa S, Jernigan RL (1985). Estimation of effective interresidue contact energies from protein crystal structure: quasi-chemical approximations. *Macromolecules* **18**: 534-552

[33] Grantham R (1974). Amino acid difference formula to help explain protein evolution. *Science* **185**: 862-864

[34] Jones D (1975). Amino acid properties and side chain orientation in proteins: a cross correlation approach. *J. Theor. Biol.* **50**: 167-183.

[35] Uversky VN (2002). Natively unfolded proteins: a point where biology waits for physics. *Protein Sci.* **11**: 739-756.

[36] Gandy S (2005). The role of cerebral amyloid beta accumultaion in common forms of Alzheimer disease. *J. Clin. Invest.* **115**: 1121-1129.

[37] Valerio MC, Colosimo A, Conti F, Giuliani A, Grottesi A, Manetti C, Zbilut JP (2005). Early events in protein aggregation: molecular flexibility and hydrophobicity/charge interaction in amyloid peptides as studied by molecular dynamics simulation. *Proteins* **58**: 110-118.

[38] Zbilut JP (2004). *Unstable Singularities and Randomness: Their Importance in the Complexity of Physical, Biological and Social Sciences*. Amsterdam: Elsevier.

[39] Zbilut JP, Scheibel T, Huemmerich D, Webber Jr CL, Colafranceschi M, Giuliani A (2005). Statistical approaches for investigating silk properties. *Applied Physics A* DOI: 10.1007/s00339-005-3429-4

[40] Zbilut JP, Scheibel T, Huemmerich D, Webber Jr CL, Colafranceschi M, Giuliani A (2005). Spatial stochastic resonance in protein hydrophobicity. *Phys. Lett. A* **346**: 33-41.

[41] Zbilut JP, Mitchell JC, Giuliani A, Marwan N, Webber Jr CL (2004). Singular hydrophobicity patterns and net charge: A mesoscopic principle for protein aggregation/folding. *Physica A* **343**: 348–358.

[42] Lio P, Vannucci M 2000. *Bioinformatics* **16**, 376

In: Protein Folding-Misfolding...
Editors: J.P. Zbilut and T. Scheibel, pp. 163-187
ISBN: 1-60021-417-7


*Chapter 7*

# FUNDAMENTALS OF ELECTROSTATICS

***Dirk Gillespie***
Department of Molecular Biophysics and Physiology, Rush University Medical Center
Chicago, IL 60612

## Abstract

Since charges are ubiquitous in biology and electrical forces are very strong, an accurate physical understanding of a biophysical system requires an accurate and complete understanding of electrostatics. This chapter describes the fundamentals of studying electrical charges and the forces acting on them through the electric field. Advanced topics of how charge distributions of mobile particles and the electrical potential are coupled are also discussed.

## Introduction

Charges play a critical role in all of protein biology. For example,

- more than a quarter of all amino acids are charged at physiological pH [15]
- proteins exist in ionic solutions and their function is often sensitive to the concentration and types of ions in the solution
- many proteins denature when there is too low or too high a concentration of ions in the surrounding solution [15]
- protein/protein docking sites have large numbers of charged amino acids, one protein with positive charges and the complementary protein with negative charges [1]
- many proteins are activated by ions like the second messenger $Ca^{2+}$ [1], $K^+$ (e.g., pyruvate kinase [5]), and $Na^+$ (e.g., thrombin [16]).

Therefore it is important that the effects of charges during protein processes (e.g., protein folding) be properly included. The focus of this chapter is to introduce the foundations of electrostatics, the study of charges and electric fields when the charges do not move quickly enough to produce magnetic fields. This chapter is meant only to serve as an introduction to the subject and not an in-depth textbook. Several excellent textbooks exist that cover both electrostatics and electrodynamics, including the classic one by Jackson [11] and the more accessible ones by Griffiths [10] and Saslow [13].

# Charges

## Ions

At the subatomic scale, all charges are the result of positively-charged protons in the nucleus and negatively-charged electrons in orbitals around the nucleus. The total net charge of a particle is the difference in the number of protons and electrons. The simplest example of a charge is an *ion*, a molecule having more (or less) electrons than protons, giving the molecule an overall negative (or positive) charge. When a salt crystal of NaCl (table salt)—which is overall *electroneutral* (having no net charge)—dissolves in water into its constitutive elements (Na and Cl), the electrons are distributed asymmetrically so that many Na and Cl ions are produced:

$$\text{NaCl crystal} + \text{H}_2\text{O} \rightarrow \left(\text{Na}^+ + \text{Cl}^-\right)_n + \text{H}_2\text{O}$$

where the $n$ $Na^+$ and $n$ $Cl^-$ ions made up the NaCl crystal. The other atomic ions (e.g., $K^+$, $Ca^{2+}$, and $Mg^{2+}$) present in the body are produced the same way.

In proteins, "charged" amino acids (e.g., glutamate, aspartate, lysine, arginine) are only charged if they are in the correct local environment. Each of these amino acids has ionizable chemical groups that can gain or loose a proton ($H^+$) to give the amino acid an overall net charge [15]. The proton binding reaction depends on the *local* pH of the solution near the amino acids. Since protons are positive ions, they are attracted to negative charges and repelled by positive charges; thus, the local pH depends on how many other charged amino acids are nearby.

In the SI (MKS) system of units, charge is expressed in units of *Coulombs* (abbreviated *C*). The *fundamental charge* $e_0$ is that of a proton and is $1.60217733\times10^{-19}$ Coulombs. An electron has charge $-e_0$. The charge of an ion is usually expressed in terms of its *valence*, the number of fundamental charges it has. For example, $Na^+$ has charge $+e_0$ and therefore has valence +1. An ion with valence +1 or –1 is called *monovalent.* Similarly, $Ca^{2+}$ has charge $+2e_0$ and has valence +2. An ion with valence +2 or –2 is called *divalent.*

## Partial Charges

The total charge of a molecule is always an integer number of the fundamental charge $e_0$ (the total charge the difference in the number of protons and electrons), but there may be a partial net charge *locally* on a molecule. In covalent bonds, the outer electrons of the atoms are shared by all the nearby nuclei and, depending on the local structure of the atoms that make up the molecule, those electrons are nonuniformly distributed over the molecule's surface; the probability of finding on electron in location A can be much higher than in location B. One important example of this effect is water where electrons are more likely to be found near the oxygen than near the hydrogens, giving the oxygen end a net $-0.7e_0$ charge (approximately) and each hydrogen a net $+0.35e_0$ charge (approximately). Overall, water ($H_2O$) is electroneutral, but locally there is an asymmetric distribution of charge (**Fig. 1A**). The highly-asymmetric charge distribution of water means that water interacts strongly with nearby charges, including those from other water molecules. Most chemistry occurs because molecules are close to each other, and therefore most chemical properties of water depend on its asymmetric charge distribution.

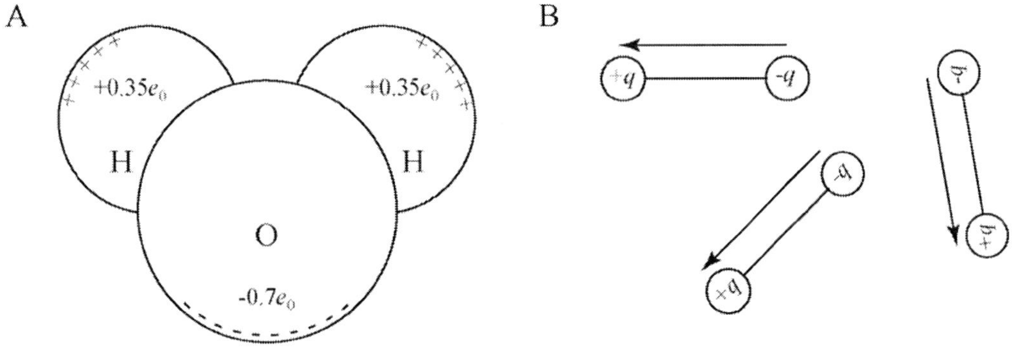


Figure 1. (A) While a water ($H_2O$) molecule is overall charge-neutral, the electrons are asymmetrically distributed (on average). Therefore, there is a net negative charge on the oxygen atom and a net positive charge on each hydrogen. (B) Dipoles in different orientations. Each dipole is represented as a dumbbell with a positive charge on one end and a negative charge on the other that are always separated by the same distance. Next to each dipole is its dipole moment, the vector that goes from the center of the negative charge to the center of the positive charge.

Other examples of this phenomenon include the charge distribution seen in crystal structures of proteins and the *dipole*. A dipole is a separation of two (partial) charges on a rigid structure that can move (but the two charges are always the same distance apart) (**Fig. 1B**). The vector that points from the center of the negative charge to the center of the positive charge is called the *dipole moment*.

## Induced Charges

The kinds of charges discussed so far are inherent to the molecule; that is, they are present when the molecule is situated in vacuum with nothing else around. Now consider a negative charge (e.g., an electron) that is fixed in space near an atom. Because the charge is negative, it will attract the nucleus and will repel the atom's electrons (**Fig. 2**). That is, the local charge structure of the atom is changed by the nearby charge (**Fig. 2**). Furthermore, the closer the charge moves to the atom, the more the atom's charge distribution will change so that the net charge distribution within the atom depends on the location of the nearby charge; the nearby charge *induces* a charge distribution that is a function of the location of the *source charge*, the charge that produces the induced charged.

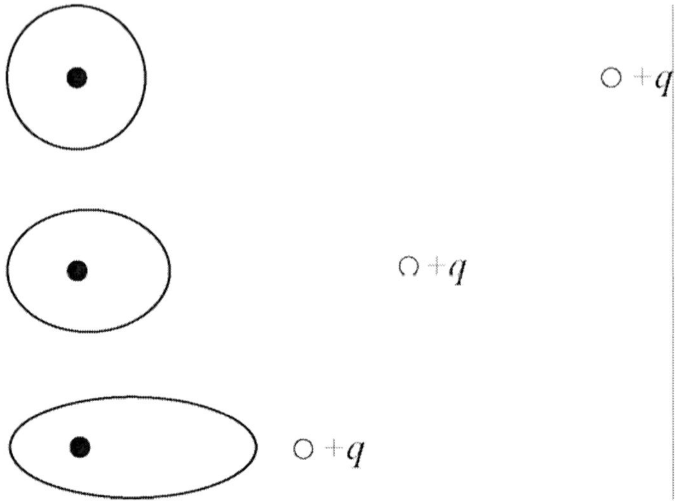


Figure 2. As a positive charge (hollow circle) approaches an atom, the nucleus (solid circle) is repelled and the electrons are attracted, distorting their orbital (line).

## Mathematics of Charge Distributions

While one can visualize charge distributions of proteins in their crystal structures, a mathematical representation of the distribution is necessary when one wants to quantify the forces on charged particles (e.g., ions, specific amino acids, or entire protein subunits). Suppose that there is only one discrete, isolated charge in vacuum and assume for now that all its charge is located at a single point. This is referred to a *point charge*. An electron is a real example of a point charge; a radius for an electron has not been measured. Later, the mathematics of electrostatics will show that many other kinds of "more realistic" charge distributions can be represented *mathematically* as collections of point charges of different signs at different locations.

On the other hand, a charge distribution may be *continuous* and not made of a finite number of point charges at particular locations. For example, the charge distribution on a protein or water molecule is continuous because of their distorted electron orbitals. Another kind of continuous charge distribution is the distribution of $Na^+$ ions in a glass of water with NaCl dissolved in it. Looking at the glass, the $Na^+$ ions are at the same concentration everywhere; their distribution is continuous. But, ions have discrete charge ($+e_0$ for $Na^+$ and $-e_0$ for $Cl^-$) and in a "snapshot" of the salt-water solution at the atomic scale their distribution is certainly *not* continuous. The continuous distribution one sees when looking on the length scale of the glass is the result of averaging many individual atomic-scale *discrete* distributions (snapshots) is an average. A temporal average is simply the time average (i.e., the integral over time) of the spatial distribution. The temporal average is observed more or less directly in a picture if the shutter is left open for a long time. One can also average abstractly by varying some property of the system (e.g., initial position of atoms) and adding up all the variants. This is called an *ensemble average* [3]. The details of this averaging are beyond the scope of this chapter, but, when calculating the electrostatics of a system of atoms, it is important to remember, as will be shown later, that there are effects due solely to the discreteness of charge and that *cannot* be calculated from the averaged, continuous charge distribution.

Mathematically, it is most convenient to represent a continuous charge distribution as a continuous *charge density,* the number of charges per volume. In the salt-water example above, the concentration of $Na^+$ is, for example, 100 millimolar (mM) which has *number density* (number of particles per volume)

$$\begin{aligned} 100\text{mM} &= 0.1\frac{\text{moles}}{\text{liter}} = 0.1N_A\frac{\text{particles}}{\text{liter}} \\ &= 100N_A\frac{\text{particles}}{\text{m}^3} = 6.022\times10^{25}\ \text{m}^{-3}. \end{aligned} \tag{1}$$

Here, $N_A$ is Avogadro's number. Since each $Na^+$ has $+e_0$ charge, the charge density of $Na^+$ is

$$e_0\cdot 6.022\times10^{25}\ \text{m}^{-3} = 9.648\times10^{6}\ \text{C/m}^3. \tag{2}$$

Similarly, the $Cl^-$ charge density is $-9.648\times10^{6}\ \text{C/m}^3$.

If $\rho(x,y,z)$ is the function that describes the charge density at each point $\mathbf{x}=(x,y,z)$, then the total charge $q$ inside a volume $V$ is given by the integral of $\rho$ over $V$:

$$q = \int_V \rho(x,y,z)\,dxdydz. \tag{3}$$

This can be seen by considering a small box with sides of length $\Delta x$, $\Delta y$, and $\Delta z$. Because $\rho(x,y,z)$ is the amount of charge per volume, the total charge $\Delta q$ near $\mathbf{x}=(x,y,z)$ is $\Delta q = \rho(x,y,z)\Delta x\Delta y\Delta z$. Then the total charge in the volume $V$ is

$$q = \sum_{\mathbf{x}\in V}\Delta q(\mathbf{x}) = \sum_{\mathbf{x}\in V}\rho(\mathbf{x})\Delta x\Delta y\Delta z \tag{4}$$

which becomes Eq. (3) as $\Delta x$, $\Delta y$, and $\Delta z$ become small.

When the charge distribution is not throughout a material, but rather on the surface or edge of material, the distribution is called a *surface* or *line charge,* respectively. Similar to the charge density of three-dimensional space, a surface charge is expressed as the number of charges per area and a line charge as the number of charges per distance. The total charge $q$ in on a surface $S$ with surface charge density $\sigma(\mathbf{s})$ (with $\mathbf{s}$ on the surface $S$) is given by the surface integral

$$q = \int_S \sigma(\mathbf{s})\, dA. \tag{5}$$

Similarly to Eq. (4), one divides the surface $S$ into small patches with area $\Delta A$ and charge $\sigma(\mathbf{s})\Delta A$ and then adding up these patches. In the limit of infinitely small patches, the sum becomes an integral.

## Forces on Charges

### Coulomb's Law

Suppose that two point charges (with charge $q_1$ and $Q$) are located at $\mathbf{r}_1$ and $\mathbf{r}$. We know that oppositely charged particles attract each other and like-charged particles repel each other. How much they attract or repel each other is the force between them. Based on experiments, the force $\mathbf{F}$ on $Q$ exerted by $q_1$ is given by Coulomb's law:

$$\mathbf{F} = \frac{Qq_1}{4\pi\varepsilon_0} \frac{\mathbf{r} - \mathbf{r}_1}{\left|\mathbf{r} - \mathbf{r}_1\right|^3}. \tag{6}$$

Here $\varepsilon_0$ is a physical constant, the *permittivity of free space*. In SI units,

$$\varepsilon_0 = 8.854 \times 10^{-12} \; \frac{\mathrm{C}^2}{\mathrm{N} \cdot \mathrm{m}^2}. \tag{7}$$

In plain words, Coulomb's law states several things:

1. The force between two charges is proportional to the product of their charges. This implies that the force between two divalent charges is four times stronger than the force between two monovalent charges the same distance apart.
2. $Q$ is pushed or pulled along a straight line connecting the two charges; the vector $\mathbf{r} - \mathbf{r}_1$ has its head at $Q$ (the charge being acted on) and its tail at $q_1$ (the charge exerting the force). When $Q$ and $q_1$ have the same sign ($Qq_1 > 0$), then $Q$ is pushed away from $q_1$ in the direction $\mathbf{r} - \mathbf{r}_1$. When they have opposite signs ($Qq_1 < 0$), $Q$ is pulled toward $q_1$ in the opposite direction.
3. The force on $q_1$ exerted by $Q$ (the opposite of what we just considered) is in exactly the opposite direction of the force on $Q$ exerted by $q_1$.
4. The magnitude of the force $|\mathbf{F}|$ drops off with the square of the distance between the charges: $\left|\mathbf{F}\right| \propto 1/r^2$ where $r = \left|\mathbf{r} - \mathbf{r}_1\right|$ is the distance between the charges.
5. No other variables are needed to describe the electric force. Coulomb's law is all that is needed and it is precise, one of the most precise and general physical laws known.

## Superposition and the Electric Field

Now suppose that there several point charges $q_1, q_2, \ldots, q_n$ at locations $\mathbf{r}_1, \mathbf{r}_2, \ldots, \mathbf{r}_n$, respectively. Experiments show that the force $\mathbf{F}$ on a charge $Q$ at $\mathbf{r}$ is the sum of the individual forces on $Q$:

$$\begin{aligned}\mathbf{F} &= \frac{Qq_1}{4\pi\varepsilon_0}\frac{\mathbf{r}-\mathbf{r}_1}{\left|\mathbf{r}-\mathbf{r}_1\right|^3} + \frac{Qq_2}{4\pi\varepsilon_0}\frac{\mathbf{r}-\mathbf{r}_2}{\left|\mathbf{r}-\mathbf{r}_2\right|^3} + \cdots + \frac{Qq_n}{4\pi\varepsilon_0}\frac{\mathbf{r}-\mathbf{r}_n}{\left|\mathbf{r}-\mathbf{r}_n\right|^3} \\ &= \frac{Q}{4\pi\varepsilon_0}\sum_{i=1}^{n} q_i \frac{\mathbf{r}-\mathbf{r}_i}{\left|\mathbf{r}-\mathbf{r}_i\right|^3}.\end{aligned} \tag{8}$$

The additivity of the forces due to source charges is called the *principle of superposition.* Eq. (8) can also be rewritten in terms of a new quantity, the *electric field* $\mathbf{E}$:

$$\mathbf{F} = Q\mathbf{E} \tag{9}$$

where

$$\mathbf{E}(\mathbf{r}) = \frac{1}{4\pi\varepsilon_0}\sum_{i=1}^{n} q_i \frac{\mathbf{r}-\mathbf{r}_i}{\left|\mathbf{r}-\mathbf{r}_i\right|^3}. \tag{10}$$

The electric field is written as a function of $\mathbf{r}$ because the formula is valid for *any* location $\mathbf{r}$, if there is a charge there or not. It represents the force a charge of $+e_0$ would feel if it were located at $\mathbf{r}$ if there were $n$ charges $q_1, q_2, \ldots, q_n$ located at $\mathbf{r}_1, \mathbf{r}_2, \ldots, \mathbf{r}_n$.

Suppose that instead of $n$ discrete charges, there is a continuous charge distribution $\rho(\mathbf{x})$. Then the same arguments that led to Eq. (8) still hold, except that at every location $\mathbf{x}$ there is an infinitesimal charge $\rho(\mathbf{x})dV$ in the infinitesimal volume $dV$ around $\mathbf{x}$. The sum in the definition of the electric field then becomes an integral:

$$\mathbf{E}(\mathbf{r}) = \frac{1}{4\pi\varepsilon_0}\int \rho(\mathbf{x})\frac{\mathbf{r}-\mathbf{x}}{\left|\mathbf{r}-\mathbf{x}\right|^3}dV. \tag{11}$$

($\mathbf{E}$ is a vector so that the integral over the vector $\mathbf{r}-\mathbf{x}$ means taking the integral of each component of $\mathbf{r}-\mathbf{x}$. When no limits of integration are given, it is implied that the integral is over all space. In Cartesian coordinates, $dV = dxdydz.$ In spherical coordinates, $dV = r^2\sin(\phi)\,drd\phi d\theta.$) One can recover Eq. (10) from Eq. (11) by defining a point charge as a Dirac delta function $\delta(\mathbf{x})$ [6]:

$$\delta(\mathbf{x}) = \begin{cases}\infty & \text{if } \mathbf{x} = 0 \\ 0 & \text{otherwise.}\end{cases} \tag{12}$$

Defining

$$\rho(\mathbf{x}) = \sum_{i=1}^{n} \delta(\mathbf{x} - \mathbf{r}_i) \tag{13}$$

and using the relation of delta functions that

$$f(\mathbf{x}) = \int f(\mathbf{x}') \delta(\mathbf{x} - \mathbf{x}') d\mathbf{x}' \tag{14}$$

Eq. (10) is recovered.

# Exploring the Electric Field

## Some Properties of the Electric Field

For a single point charge $q$ at the origin, the electric field at any location $\mathbf{r}$ is

$$\mathbf{E}(\mathbf{r}) = \frac{q}{4\pi\varepsilon_0} \frac{\mathbf{r}}{|\mathbf{r}|^3}. \tag{15}$$

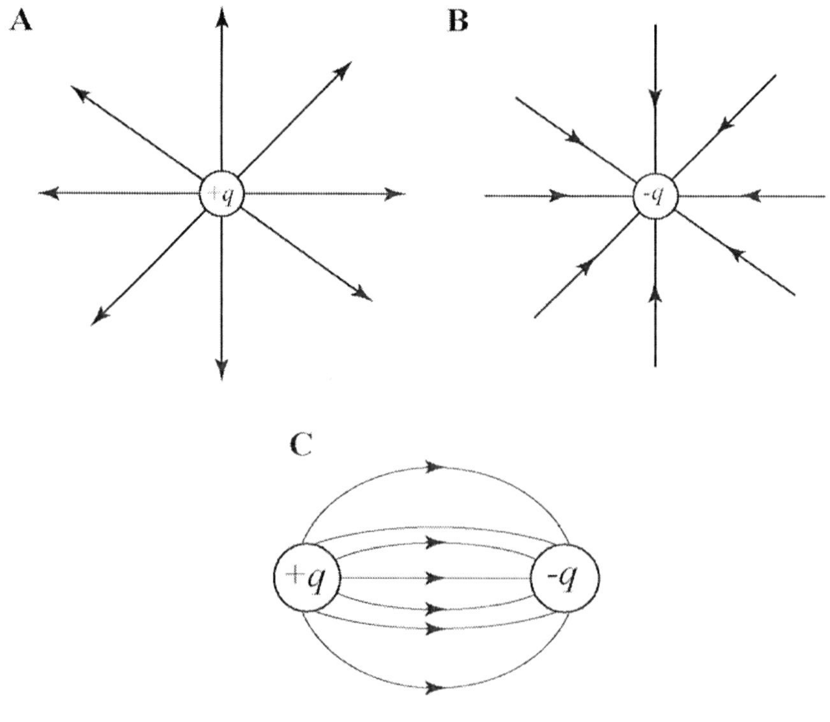


Figure 3. Field lines, by convention, start at positive charges and end at negative charges (or go out to infinity).

That is, **E** is a vector field with the *field lines*, by convention, starting at a positive charge and ending at a negative charge **(Fig. 3)**. This means that the field lines from a single point charge are straight and therefore they do not intersect.

For more than one source charge, the electric field is the sum of all the individual ones from each source charge (like in Eq. (10)). Except at the locations of these point charges ($\mathbf{r}_1, \mathbf{r}_2, \ldots, \mathbf{r}_n$), **E** is well-defined; it is always possible to add up these vectors and get a single, finite number for each component of the vector **E**. This means that even when there are multiple source charges, field lines never intersect; if they did, then at the point of intersection the vector **E** would point in two different directions at the same time.

## Counting Field Lines and Gauss's Law

Suppose that there is a sphere centered at the origin with radius $R$ and that there is a single point charge at the origin. If one were to try and count the number of field lines that pass through the sphere, there would of course be an infinite number of them. But, suppose for the moment that there is a finite number (call it $N$) of field lines that point outward from (for a charge with valence +1) or inward to (for a charge with valence −1); if the charge has valence $\pm z$, then $zN$ field lines would point out from or into it. Then it is easy to see that no matter how big or small the radius of the sphere $R$ is that the number of field lines passing through the sphere is the same and that this number is determined by the valence of the charge at the origin (**Fig. 4A**). By the same logic, the same result is true if the charge is located *anywhere inside* the sphere (**Fig. 4B**).

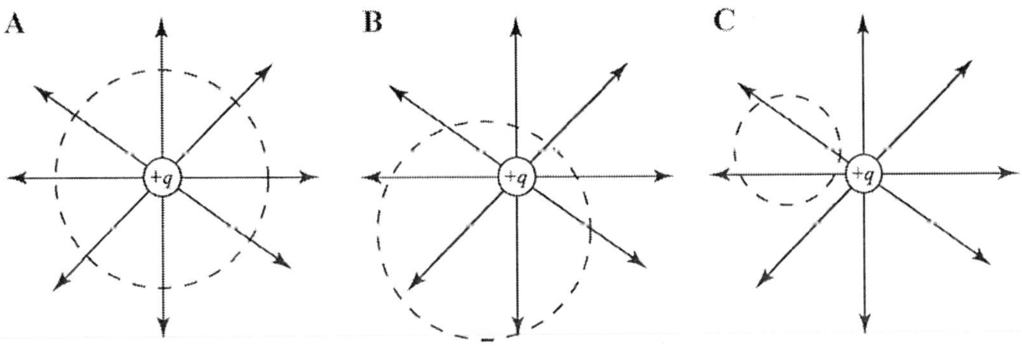


Figure 4. Counting a field line as +1 when it leaves the sphere (dashed line) and as −1 when it leaves the sphere, the net flux of the field line is established.

On the other hand, suppose that the sphere is now located so that the charge at the origin is not inside the sphere. Then, the field lines from the charge either do not enter the sphere, or they enter at one point, but then exit the sphere at another point (**Fig. 4C**). Either way, the *net* number of field lines passing through the sphere is exactly 0 by counting any field line going from inside the sphere out as positive and one going from outside to the inside as negative.

This is the essence of Gauss's Law: "the number of field lines" going through any closed surface (not just a sphere) is directly related to the number of charges inside the surface (even if there are several charges or a continuous charge distribution inside). The surface does *not* have to be a real object; any hypothetical surface (called a *Gaussian surface*) will do.

As described, this version of Gauss's Law is not true because there are an infinite number of field lines that emanate from a charge. To define "counting field lines" properly, we need to define the *flux* of **E** through a surface $S$:

$$\Phi_E = \int_S \mathbf{E} \cdot \mathbf{n} dS \tag{16}$$

where $\mathbf{n}(\mathbf{s})$ is the unit outward normal vector at $\mathbf{s}$ (i.e., the vector of length 1 that points perpendicularly outward from the surface at $\mathbf{s}$) on the surface $S$ and $\mathbf{E}(\mathbf{s}) \cdot \mathbf{n}(\mathbf{s})$ is the component of the electric field that is perpendicular to the surface $S$ at the point $\mathbf{s}$. $dS$ is an infinitesimal patch of surface so that the surface area of $S$ is given by $\int_S dS$ [12]. Then the flux is a measure of how much of the electric field is going through $S$.

Using the definition of the electric field, let's go back to the case of one point charge $q$ at the origin and $S$ being a sphere of radius $R$ centered at the origin and calculate the flux of **E** through $S$. For $S$, at any point $\mathbf{s}$ on the surface the unit normal is $\mathbf{s}/R$. By Eq. (15), the flux of **E** is

$$\begin{aligned}
\int_S \mathbf{E} \cdot \mathbf{n} dS &= \int_S \frac{q}{4\pi\varepsilon_0} \frac{\mathbf{s}}{|\mathbf{s}|^3} \cdot \frac{\mathbf{s}}{R} dS \\
&= \frac{q}{4\pi\varepsilon_0} \frac{1}{R^4} \int_S \mathbf{s} \cdot \mathbf{s} dS \\
&= \frac{q}{4\pi\varepsilon_0} \frac{1}{R^2} \int_S dS \\
&= \frac{q}{\varepsilon_0}
\end{aligned} \tag{17}$$

where we used the fact that $\mathbf{s}$ is on the sphere $S$ and therefore $|\mathbf{s}| = R$ and that $\mathbf{s} \cdot \mathbf{s} = |\mathbf{s}|^2 = R^2$. The result that

$$\int_S \mathbf{E} \cdot \mathbf{n} dS = \frac{q}{\varepsilon_0} \tag{18}$$

where $q$ is total net charge inside $S$ is *Gauss's Law* and is true, like the intuitive version before, for *all* closed surfaces (not just spheres) and for many charges (or even continuous charge distributions) inside the surface. Again, $S$ does not have to be a real object, but any hypothetical surface as long as it is closed so no field lines can escape without being counted.

Overall, Gauss's Law is an important starting point, both for solving problems and for other important ideas in electrostatics.

**Gauss's Law and Symmetry**

The integral in Gauss's Law, like all integrals, only tells you something about average properties of the integrand. For example, in Gauss's Law, the integral tells you about the total flux of **E**, but it does not tell you about the distribution of field lines on the *S* (i.e., if they are at a higher density in one spot on *S* than at another spot). One of the practical applications of Gauss's Law is when you know before starting the problem that the density of field lines is the same all over *S*, usually because the problem has symmetry (e.g., it is spherically symmetric and therefore all functions only depend on how far away from the origin one is, not on the angle made with respect to the origin) and because the Gaussian surface has been chosen carefully.

To illustrate the point, consider a solid sphere *B* of radius *b* at the origin with a uniform continuous charge distribution $\rho$ and total charge $q = \rho 4\pi b^3 / 3$. Because the sphere has a uniform charge distribution, we know that the electric field coming out of the sphere will be spherically symmetric. If one is clever about picking a Gaussian surface, then one can exploit this symmetry to calculate the electric field anywhere outside *B*. Because of the symmetry, the Gaussian surface *S* should have the same symmetry so the natural choice is sphere centered at the origin with radius $r > b$.

Let **r** be on *S* so that $|\mathbf{r}| = r$. Because of the spherical symmetry, the electric field at any location is vector that points in the direction of **r**:

$$\mathbf{E}(\mathbf{r}) = E(r)\frac{\mathbf{r}}{|\mathbf{r}|}. \tag{19}$$

where $E(r) = |\mathbf{E}(\mathbf{r})|$. Therefore,

$$\mathbf{E}(\mathbf{r})\cdot\mathbf{n}(\mathbf{r}) = E(r)\frac{\mathbf{r}}{|\mathbf{r}|}\cdot\frac{\mathbf{r}}{|\mathbf{r}|} = E(r) \tag{20}$$

and so

$$\int_S \mathbf{E}\cdot\mathbf{n}dS = E(r)\int_S dS = 4\pi r^2 E(r). \tag{21}$$

By Gauss's Law,

$$4\pi r^2 E(r) = \frac{q}{\varepsilon_0} \tag{22}$$

or

$$E(r) = \frac{q}{4\pi\varepsilon_0 r^2}. \tag{23}$$

Then,

$$\mathbf{E}(\mathbf{r}) = \frac{q}{4\pi\varepsilon_0 r^3}\mathbf{r}. \tag{24}$$

This simple result looks exactly like Eq. (15) with $r = |\mathbf{r}|$; the electric field outside a ball of uniformly charged material is the same as if all the charge had been put into a point charge at the origin.

**Surface Charges and the Electric Field**

Another application of Gauss's Law is showing the effect of a surface charge on the electric field. To that end, suppose that an electric field is passing through a surface with a uniform surface charge $\sigma$. To apply Gauss's Law, a Gaussian surface $S$ must be chosen. In this case, we choose a particular Gaussian surface called a *Gaussian pillbox*: a rectangular box with a top and bottom of area $A$ parallel to the plane and a very small height $\delta$ (**Fig. 5**). Assuming that the surface is relatively smooth, it is always possible to choose $A$ small enough that surface underneath is approximately flat around the point $\mathbf{r}$ on the surface that we choose.

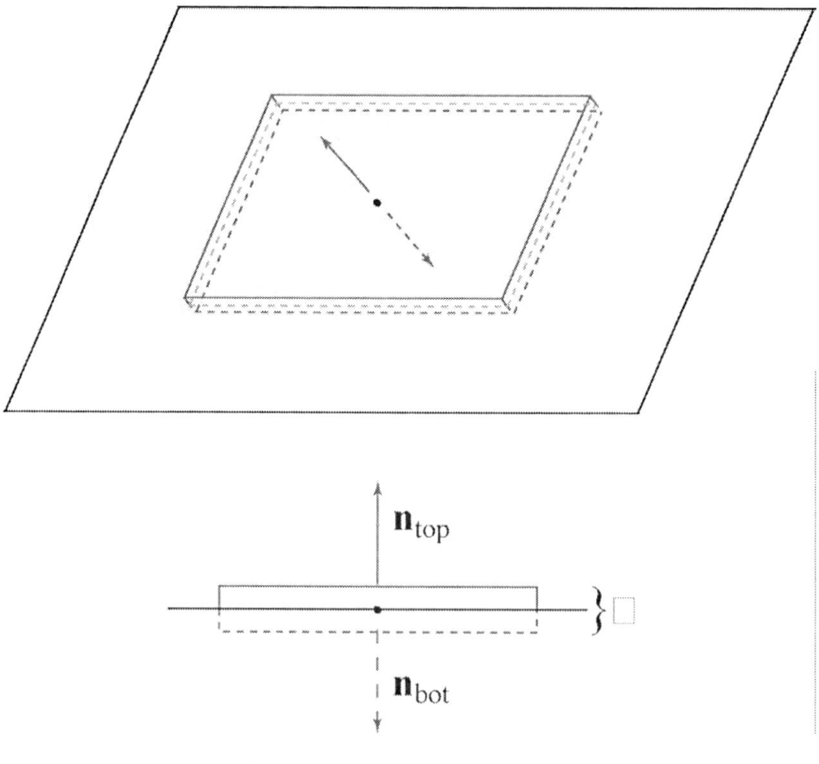


Figure 5. Gaussian pillbox (red) near flat surface (black).

Because $S$ is a box, it has two *outward* unit normals, one on the top ($\mathbf{n}_{\text{top}}$) and one on the bottom of the box ($\mathbf{n}_{\text{bot}}$). In the limit of the height of the box $\delta$ going to 0,

$$\mathbf{n}_{\text{top}} \to \mathbf{n}_{\text{surf}} \qquad \mathbf{n}_{\text{bot}} \to -\mathbf{n}_{\text{surf}} \tag{25}$$

where $\mathbf{n}_{\text{surf}}$ is a unit normal of the charged surface at $\mathbf{r}$. (This is possible because the charged surface is flat in the tiny area that we are examining with the Gaussian pillbox.) Then Gauss's Law states that

$$\int_S \mathbf{E} \cdot \mathbf{n} dS = \frac{q}{\varepsilon_0} = \frac{\sigma A}{\varepsilon_0}. \tag{26}$$

For the left-hand side, by our choice of a Gaussian pillbox with $A$ small enough so that the surface underneath is approximately flat so that the electric field is almost constant we have

$$\begin{aligned} \int_S \mathbf{E} \cdot \mathbf{n} dS &= \left(\mathbf{E}_{\text{top}} \cdot \mathbf{n}_{\text{top}} + \mathbf{E}_{\text{bot}} \cdot \mathbf{n}_{\text{bot}}\right) A + O(\delta) \\ &\to \left(\mathbf{E}_{\text{top}} \cdot \mathbf{n}_{\text{surf}} - \mathbf{E}_{\text{bot}} \cdot \mathbf{n}_{\text{surf}}\right) A \qquad \text{as } \delta \to 0. \end{aligned} \tag{27}$$

In words, the normal component *out* of the Gaussian pillbox

1. is almost entirely through the top or bottom with the sides contributing a negligible amount as $\delta$ gets smaller and smaller (the order $\delta$ term at the end)
2. is, in the limit of $\delta \to 0$, the difference in the normal components of the electric field just above and just below the charged surface.

Combining the two sides of Gauss's Law, we get

$$\mathbf{E}_{\text{top}} \cdot \mathbf{n}_{\text{surf}} - \mathbf{E}_{\text{bot}} \cdot \mathbf{n}_{\text{surf}} = \frac{\sigma}{\varepsilon_0}. \tag{28}$$

This equation is important because it states that the electric field is discontinuous across a surface with a charge; the electric field jumps at the surface. If the electric field were continuous,

$$\mathbf{E}_{\text{top}} \cdot \mathbf{n}_{\text{surf}} = \mathbf{E}_{\text{bot}} \cdot \mathbf{n}_{\text{surf}}. \tag{29}$$

Moreover, the size of the jump in the electric field is determined solely by the surface charge. Eq. (28) also says that whenever an electric field crosses a boundary that does *not* have a surface charge (i.e., $\sigma = 0$), then the electric field is continuous across that boundary.

### Differential Form of Gauss's Law

Gauss's Law (Eq. (18)) contains an integral. As described above, it only describes the net flux of the field lines and not their distribution. Unless there is some symmetry in the problem, this integral form of Gauss's Law is not as useful as the differential form that, with a little math, can be derived from the integral form.

Suppose that there is a charge distribution $\rho(\mathbf{r})$. Then at any point where $\rho$ is defined, one can put a Gaussian surface *S* around that point. Inside *S*, the charge enclosed is

$$q = \int_V \rho(\mathbf{r})\,dV \tag{30}$$

where *V* is the volume of space enclosed by *S*. The other half of Gauss's Law can be rewritten by Gauss's Theorem from calculus [12]:

$$\int_S \mathbf{E}\cdot\mathbf{n}\,dS = \int_V (\nabla\cdot\mathbf{E})\,dV. \tag{31}$$

Equating these using Gauss's Law, we get

$$\int_V \left(\nabla\cdot\mathbf{E}(\mathbf{r}) - \frac{\rho(\mathbf{r})}{\varepsilon_0}\right) dV = 0. \tag{32}$$

Now recall that at the beginning no assumptions were made about the Gaussian surface *S*; it was any arbitrary closed surface. Therefore, Eq. (32) is true for *any* volume of space *V*. In order to have the integral be 0 for any *V*, the integrand must be zero:

$$\nabla\cdot\mathbf{E}(\mathbf{r}) = \frac{\rho(\mathbf{r})}{\varepsilon_0}. \tag{33}$$

This is the differential form of Gauss's Law and is more practical than the integral form because, after the differential equation is solved, it describes the electric field **E** everywhere in space.

## The Electrical Potential

To introduce the concept of the electrical potential, we start with an easily-verifiable mathematical fact about the electric field **E** of a single point charge:

$$\nabla\times\mathbf{E} = 0; \tag{34}$$

that is, the curl of **E** is 0 so that **E** is irrotational [12]. The superposition principle then implies that the curl of the electric field resulting from *any* charge distribution is also 0:

$$\nabla \times \mathbf{E} = \nabla \times \mathbf{E}_1 + \nabla \times \mathbf{E}_2 + \nabla \times \mathbf{E}_3 + \cdots = 0 \tag{35}$$

where the $\mathbf{E}_i$ are the electric fields from charges throughout the system. From the Stokes theorem of calculus it follows that there must be a scalar (i.e., nonvector) function $\phi(\mathbf{r})$ so that

$$\mathbf{E}(\mathbf{r}) = -\nabla\phi(\mathbf{r}); \tag{36}$$

that is, the electric field (for a given charge distribution) is the (negative) gradient of the *electrical potential* $\phi$ [10-13].

The units of electrical potential are *volts* (abbreviated *V*) and the units of the electric field actually derived from it as *volts/meter*. The most easily seen difference between the electrical potential and the electric field is that $\phi$ is a scalar function (it produces one number), not a vector-valued function. In general, the electrical potential can be used to describe work and electrical energy in a system. For example, the work $W$ done to move a charge $q$ from point **a** to point **b** is

$$W = q\left(\phi(\mathbf{b}) - \phi(\mathbf{a})\right) \tag{37}$$

and the electrical energy of a system with charge distribution $\rho(\mathbf{r})$ is

$$\frac{1}{2}\int \rho(\mathbf{r})\phi(\mathbf{r})\,d\mathbf{r}. \tag{38}$$

Details of this can be found in the standard textbooks.

One of the things to keep in mind with the electrical potential is that because its *derivative* is the electric field, $\phi(\mathbf{r})$ is only defined up to an additive constant. Therefore, one only discusses electrical potential *differences*. Because it is not possible to define an absolute number for the electrical potential, potentials are always referenced to the electrical potential at some location that is defined to be *grounded* (having zero electrostatic potential). For example, in a system with isolated charges, the potential at infinity is usually chosen to be zero. In other systems, electrodes clamp the electrical potential with one electrode defined to be at zero potential and the other electrode(s) actively maintained (by external electrical circuits) at a specified electrical potential. In experiments, it is essential to have a well-defined ground potential that does not vary rapidly with respect to the world outside the experiment. This is often done by placing the experimental apparatus in a box of metal that serves as a physical manifestation of the Gaussian pill box discussed above.

## The Electrical Potential of Charge Distributions

If there are several point charges $q_1, q_2, \ldots, q_n$ at locations $\mathbf{r}_1, \mathbf{r}_2, \ldots, \mathbf{r}_n$, then it is easy to show that the electrostatic potential of each individual charge *i* is given by

$$\frac{1}{4\pi\varepsilon_0}\frac{q_i}{\left|\mathbf{r}-\mathbf{r}_i\right|} \tag{39}$$

and that the electrical potential of all the charges together is the sum of the individual electrical potentials:

$$\phi(\mathbf{r}) = \frac{1}{4\pi\varepsilon_0}\sum_{i=1}^{n}\frac{q_i}{\left|\mathbf{r}-\mathbf{r}_i\right|}. \tag{40}$$

This means that the electrical potential also follows the principle of superposition. By superposition, a charge distribution yields an electrical potential given by

$$\phi(\mathbf{r}) = \frac{1}{4\pi\varepsilon_0}\int\frac{\rho(\mathbf{r}')}{\left|\mathbf{r}-\mathbf{r}'\right|}d\mathbf{r}'. \tag{41}$$

Of course, the summation and integrals imply that the ground potentials in each of the *i* charge distributions is the same.

## The Poisson Equation

The electrical potential can also be determined from a charge distribution $\rho(\mathbf{r})$ by the differential form of Gauss's Law (after substituting in the definition of the potential):

$$-\varepsilon_0\nabla^2\phi(\mathbf{r}) = \rho(\mathbf{r}) \tag{42}$$

where the *Laplacian operator* $\nabla^2$ is

$$\nabla^2 f(\mathbf{r}) = \nabla\cdot\left(\nabla f(\mathbf{r})\right) = \frac{\partial^2 f}{\partial x^2}+\frac{\partial^2 f}{\partial y^2}+\frac{\partial^2 f}{\partial z^2} \tag{43}$$

with $\mathbf{r} = (x, y, z)$.

The *Poisson equation* (as Eq. (42) is called) is a partial differential equation unless there is enough symmetry in the system being studied so that both the electrical potential and the charge distribution are functions of only one variable (the Poisson equation is then an ordinary differential equation). Of course, when there is only one spatial variable, it is much easier to solve the Poisson equation than if there are three spatial variables. In fact, how to most efficiently and accurately solve the general Poisson equation (i.e., for arbitrary charge distributions $\rho(\mathbf{r})$) is still an active field of research.

To solve the Poisson equation (or *any* differential equation) one needs to know more than just $\rho(\mathbf{r})$. The same charge distribution $\rho(\mathbf{r})$ can give different electrical potentials $\phi(\mathbf{r})$ depending on the *boundary conditions* that define what values $\phi$ (or its derivative) take on the boundaries of the system. In fact, no solution to a differential equation is uniquely defined unless boundary conditions are specified. If the differential equation is a partial differential equation, the need for boundary conditions is particularly clear: different boundary conditions add different functions to the solution. (In ordinary differential equations, different boundary conditions add different constants to the solution.) The physical meaning of the boundary conditions is clear in experimental and engineering applications. The boundary conditions specify what is injected into the system. It is obvious in those applications that isolated systems (with no injection of anything) have very different properties from driven systems.

To see the role of boundary conditions, consider a system with an infinite plane of uniform surface charge $\sigma$ so that the system has planar symmetry and the potential changes only in the $x$ direction and $\rho(x)=\sigma\delta(x)$. Then the Poisson equation is

$$-\varepsilon_0 \frac{d^2\phi}{dx^2}=\sigma\delta(x). \tag{44}$$

Integrating this equation once gives

$$-\varepsilon_0 \frac{d\phi}{dx}+C_1=\sigma H(x) \tag{45}$$

where $H$ is the Heaviside step function:

$$\begin{aligned} H(x) &= \begin{cases} 0 & \text{if } x<0 \\ 1 & \text{if } x>0 \end{cases} \\ &= \int_{-\infty}^{x} \delta(x')dx'. \end{aligned} \tag{46}$$

The constant $C_1$ is a constant of integration. Eq. (45) is consistent with the result we have already found for the electric field (Eq. (28)):

$$\begin{aligned} \mathbf{E}_{\text{top}}\cdot\mathbf{n}_{\text{surf}} &= -\frac{d\phi}{dx}(x\to 0; x>0)=\frac{\sigma-C_1}{\varepsilon_0} \\ \mathbf{E}_{\text{bot}}\cdot\mathbf{n}_{\text{surf}} &= -\frac{d\phi}{dx}(x\to 0; x<0)=-\frac{C_1}{\varepsilon_0}. \end{aligned} \tag{47}$$

Integrating Eq. (45) to find $\phi$, we get

$$-\varepsilon_0\phi(x)+C_1x+C_2=\begin{cases}0 & \text{if } x<0\\ \sigma x & \text{if } x\geq 0\end{cases} \tag{48}$$

where $C_2$ is another constant of integration. To determine the two unknowns $C_1$ and $C_2$ we need to know something about $\phi$ somewhere (i.e., the boundary conditions). A common simplification that is made is to assume that boundary conditions are applied at infinity (in this case at $x=-\infty$ and $x=+\infty$), but it should be immediately obvious that this is impossible in this case because, for any choice of $C_1$ and $C_2$, $\phi$ is infinite at $x=+\infty$. In this case, a poorly-chosen simplification makes the different equation unsolvable because it does not have a solution for boundary conditions applied at infinity. If, however, we had two electrodes a *finite* distance from the plane of charge (even if they are very, very, very, very far away—but not infinitely far away), then the Poisson equation has a solution. If two electrodes (there is no such thing as one electrode, but there may be more than two) are located at $x=-a$ and at $x=a$ with the one at $x=-a$ maintained at voltage $V$ (i.e., $\phi(-a)=V$) and the one at $x=-a$ grounded (i.e., $\phi(a)=0$), then we can find the constants of integration:

$$\begin{aligned}0&=-\varepsilon_0\phi(-a)-C_1a+C_2=-\varepsilon_0V-C_1a+C_2\\ \sigma a&=-\varepsilon_0\phi(a)+C_1a+C_2=C_1a+C_2\end{aligned} \tag{49}$$

so that in Eq. (48)

$$\begin{aligned}C_1&=\frac{1}{2a}(\sigma a-\varepsilon_0V)\\ C_2&=\frac{1}{2}(\sigma a+\varepsilon_0V).\end{aligned} \tag{50}$$

For these two electrodes, the electrical potential $\phi$ is then completely determined.

It is possible to choose other boundary conditions (e.g., put left electrode at $x=-2a$ or require that the electric field—not the potential—has specified values at $x=-a$ and at $x=a$) and those would change the electrical potential profile throughout the system. In short, *an electrostatic potential profile cannot be determined unless boundary conditions are specified* and specifying those conditions can be subtle and difficult. If boundary conditions are not clearly stated in a calculation of the electric field, it is necessary to reconstruct the boundary conditions used in the calculation before the results can be used in confidence. The implied boundary conditions may not describe the physical system of interest.

## But the Real World Is not So Easy

So far the discussions of charge distributions, electric fields, and electrical potentials were somewhat abstract and did not touch on real-world imperfections and difficulties; the discussion so far apply to a given, unchanging charge distribution in vacuum. But, even on the atomic scale biology does not occur in vacuum. Proteins are surrounded by ions (both large—other proteins—and small—atomic ions like $K^+$ and $Ca^{2+}$) and these change the charge distribution on the protein itself, both by inducing charges and by moving larger charged subunits of the protein.

The general problem one faces in this case is that the location of moveable charges and the electric field are *nonlinearly* coupled: the charges make the electric field, but the charges move in response to the electric field. Mathematically, this means that the charge distribution $\rho(\mathbf{r})$ are location $\mathbf{r}$ is a function of the local electric field $\mathbf{E}(\mathbf{r})$ and/or the local electrical potential $\phi(\mathbf{r})$:

$$\rho(\mathbf{r}) = \rho(\mathbf{r};\mathbf{E}(\mathbf{r}),\phi(\mathbf{r})). \tag{51}$$

How $\rho$ depends on $\mathbf{E}$ and $\phi$ is not obvious, and, in addition, the electric field and electrical potential depend on the location of all other charges. Indeed, a major topic of research in all physics (from nuclear to macroscopic perhaps even astronomical) is to determine how $\rho(\mathbf{r};\mathbf{E}(\mathbf{r}),\phi(\mathbf{r}))$ depends on $\mathbf{r}$, $\mathbf{E}(\mathbf{r})$, and $\phi(\mathbf{r})$.

Next, we examine two theories—*not* laws of nature—about how $\rho$ depends on $\mathbf{E}$ and $\phi$ for two different kinds of charge distributions. Then, we will see that when these two theories are combined (which in principle they can be) the result gives physically unrealistic results.

### Linear Dielectrics

As described earlier, a source charge can induce charges in a material by attracting (or repelling) the electrons of the material's atoms while at the same time repelling (or attracting) the nuclei. Each atom is then a small dipole with the negative charge closer to a positive source charge and the positive charge farther away. In other words, the dipole moment of the atom is pointing in the direction of the electric field of the source charge (**Fig. 1B**).

Now suppose we look at our material on a larger length scale where each small part of the material is small enough to be microscopic, but large enough not to resolve individual atoms. For example, if one considers a nanoliter (one billionth of a liter) of a salt water solution, then it makes up a very, very, very tiny portion of a glass of the solution, but it still contains a huge number of atoms. If on this scale, the induced dipole moment of a small piece of material $\mathbf{P}(\mathbf{r})$ at location $\mathbf{r}$ is still proportional to the local electric field $\mathbf{E}(\mathbf{r})$ of the source charge, then the material is called a *linear dielectric material*:

$$\mathbf{P}(\mathbf{r}) = \chi\varepsilon_0\mathbf{E}(\mathbf{r}) \tag{52}$$

where $\chi$ is the *electric susceptibility* of the material (an inherent property of the material). In general, it is possible that the susceptibility is a function of location $\mathbf{r}$ (e.g., when two materials with different susceptibilities are brought into contact). Because $\mathbf{P}$ is the dipole moment of an infinitesimal volume of material, it is possible to show that the induced charge distribution $\rho_{\text{ind}}(\mathbf{r})$ is given by [10]

$$\rho_{\text{ind}}(\mathbf{r}) = -\nabla \cdot \mathbf{P}(\mathbf{r}) = -\varepsilon_0 \nabla \cdot (\chi(\mathbf{r})\mathbf{E}(\mathbf{r})). \tag{53}$$

When a charge distribution can be written as the divergence of a vector field (note most charge distributions cannot be written this way), it has special properties. For example, the charge distribution is zero in regions where $\mathbf{P}$ is constant, and charge appears at boundaries when $\mathbf{P}$ has one value on one side of the boundary and another value on the other side. This is one example of how the charge distribution can depend on the local electric field (Eq. (51)).

So what electric field $\mathbf{E}$ is this? Certainly one component must be the electric field from the distribution of *source* charges $\rho_{\text{src}}(\mathbf{r})$:

$$\varepsilon_0 \nabla \cdot \mathbf{E}_{\text{src}}(\mathbf{r}) = \rho_{\text{src}}(\mathbf{r}). \tag{54}$$

Also, the field generated by the induced charges (produced by the source charges) will be part of this electric field; $\mathbf{E}$ is the *total* electric field from both the source and induced charges. Since both the source and the induced charges are present in the system, by the principle of superposition the *total* electric field $\mathbf{E}_{\text{tot}}$ from the two together can be computed:

$$\begin{aligned} \varepsilon_0 \nabla \cdot (\mathbf{E}_{\text{tot}}(\mathbf{r})) &= \rho_{\text{tot}}(\mathbf{r}) = \rho_{\text{src}}(\mathbf{r}) + \rho_{\text{ind}}(\mathbf{r}) \\ &= \rho_{\text{src}}(\mathbf{r}) - \varepsilon_0 \nabla \cdot (\chi(\mathbf{r})\mathbf{E}_{\text{tot}}(\mathbf{r})) \end{aligned} \tag{55}$$

or

$$\varepsilon_0 \nabla \cdot (\varepsilon(\mathbf{r})\mathbf{E}_{\text{tot}}(\mathbf{r})) = \rho_{\text{src}}(\mathbf{r}) \tag{56}$$

where the *relative dielectric coefficient* of the material at $\mathbf{r}$ is

$$\varepsilon(\mathbf{r}) = 1 + \chi(\mathbf{r}). \tag{57}$$

The corresponding Poisson equation is

$$-\varepsilon_0 \nabla \cdot (\varepsilon(\mathbf{r}) \nabla \phi_{\text{tot}}(\mathbf{r})) = \rho_{\text{src}}(\mathbf{r}). \tag{58}$$

Eqs. (56) and (58) are only true for linear dielectric materials. It is not true if Eq. (52) is not true, but many materials fall into this category. Moreover, strictly speaking, a dielectric coefficient (or susceptibility) is a bulk property of a material, not an atomic-scale property.

Despite this, individual protein molecules are routinely modeled as linear dielectric materials with good success.

While the dielectric coefficient is a well-defined material property, but it can change when conditions change. For example, the dielectric coefficient of a KCl electrolyte solution can change from 78 in dilute conditions to 44 at saturation [2]. Moreover, the dielectric coefficient is different depending on the timescale you are interested in. For the KCl solution, the dielectric coefficients above are at the longest timescale when you see nothing changing in the glass (seconds). However, on the timescale of atomic motion (femtoseconds, $10^{-15}$ seconds) water and the other ions have not had time to rearrange itself in response to a moving ion; there is virtually no dielectric response on this timescale [2].

One of the reasons that modeling a material (e.g., a protein) as a linear dielectric is done so often is that the Poisson equation (58) is relatively easy to solve numerically. Another approach (rather than solving Eq. (58) directly) is to calculate the induced charge distribution $\rho_{\text{ind}}(\mathbf{r})$ directly. Once that is determined, the total (due to both source and induced charges) electrical potential can be computed using Eq. (41) with

$$\rho(\mathbf{r}) = \rho_{\text{src}}(\mathbf{r}) + \rho_{\text{ind}}(\mathbf{r}). \tag{59}$$

This method also has the advantage of treating all charges explicitly and in the same way, whether dielectric or not. Handling dielectric charges differently from other charges (e.g., the method of image charges [10; 11; 13]) is a common source of confusion—and error.

Another advantage of this method is that it can be more numerically efficient if $\varepsilon(\mathbf{r})$ is piecewise constant (i.e., the dielectric boundary between materials with different susceptibilities is sharp). Solving a Poisson equation like Eq. (58) must be done over the entire volume of the system; i.e., $\phi(\mathbf{r}_k)$ must be found at every point $\mathbf{r}_k$ of the three-dimensional grid that describes the system where the charges are. On the other hand, if $\varepsilon(\mathbf{r})$ is piecewise constant, then $\rho_{\text{ind}}(\mathbf{r})$ is a surface charge located on the dielectric interface. This is a two-dimensional surface that takes less points to describe than the three-dimensional grid used to solve the Poisson equation. The whole procedure is also relatively easy formulate and solve, but that is beyond the scope of this chapter. Further details can be found in Ref. [4].

## Poisson-Boltzmann Theory

Another example of how the charge distribution $\rho(\mathbf{r})$ can depend on the electrical potential is seen in how ions rearrange themselves around a fixed charge. Suppose that in a homogeneous solution of salt water a positively charged protein (for example) is introduced. Because of the protein's charge, the ions will redistribute themselves around this central molecule, with negative ions nearest to the protein because they are attracted to its positive charge. Positive ions are then attracted to the layer of negative ions, a pattern that quickly dissipates in a couple of ion layers.

One of the simplest theories to (approximately) calculate the final charge distribution of the ions around the protein is *Poisson-Boltzmann theory*. The details of the derivation are beyond the scope of this chapter (they may be found in Refs. [3] and [8]), but the theory approximates ions as point particles whose distribution is given by

$$p(\mathbf{r}) = \rho_b \exp\left(-\frac{e\phi(\mathbf{r})}{kT}\right) \tag{60}$$

and

$$n(\mathbf{r}) = \rho_b \exp\left(\frac{e\phi(\mathbf{r})}{kT}\right). \tag{61}$$

Here $k$ is the Boltzmann constant and $T$ is the absolute temperature. For simplicity we have assumed the two ion species are monovalent that have a bulk concentration of $\rho_b$ and the local concentration of the positive ions is $p(\mathbf{r})$ and the local concentration of the negative ions is $n(\mathbf{r})$. Eq. (61) reflects the fact that ions move in response to a local electrostatic potential.

On the other hand, charges (in this case, the ions and the protein charge distribution $\rho_{\text{pro}}(\mathbf{r})$) make the electrical potential $\phi(\mathbf{r})$. There are all related through the Poisson equation

$$-\varepsilon_0 \nabla \cdot \left(\varepsilon(\mathbf{r}) \nabla \phi(\mathbf{r})\right) = e_0 \rho_b \left(\exp\left(-\frac{e_0\phi(\mathbf{r})}{kT}\right) - \exp\left(\frac{e_0\phi(\mathbf{r})}{kT}\right)\right) + \rho_{\text{pro}}(\mathbf{r}) \tag{62}$$

where we have allowed the protein to have a different dielectric coefficient from the surrounding ion solution.

Eqs. (60)-(62) state that the ion distribution profiles are determined by the electrostatic potential created by *all* the ions. This is the *mean-field approximation* and it is true only in the limit that there are an infinite number of ions that have zero diameter. On the atomic scale, an ion does not act on itself; only the *other* ions exert a force on an ion. But, if there is only one electrostatic potential function that determines the ion distribution profiles (like in the Poisson-Boltzmann approach), then that potential is calculated from *all* the ions in the system and therefore must include this self-interaction. The mean-field approximation is justified in any reasonably-sized system with a large number of ions—except when the protein has a different dielectric coefficient, as we will see in the next section.

## Discreteness of Charge and the Dielectric Interface

As an ion in a high dielectric approaches a material with low dielectric coefficient, charge (with the same sign as the charge on the ion) is induced on the dielectric interface [10]. In the idealization of a linear dielectric, this induced charge appears instantaneously (i.e., without delay) whenever an ion is present and changes instantaneously whenever an ion moves. But, because this process is without delay, the electrostatic effect of the dielectric interface on the ions must be based on the locations of the ions at a specific time. Specifically, the instantaneous electrostatic potential $\psi(\mathbf{x})$ at location $\mathbf{x}$ is given by the Poisson equation

$$-\varepsilon_0 \nabla \cdot \left( \varepsilon(\mathbf{x}) \nabla \psi(\mathbf{x}) \right) = e \left( \sum_i z_i \delta(\mathbf{x} - \mathbf{y}_i) \right) \tag{63}$$

where the ions with valence $z_i = \pm 1$ are located at $\mathbf{y}_i$.

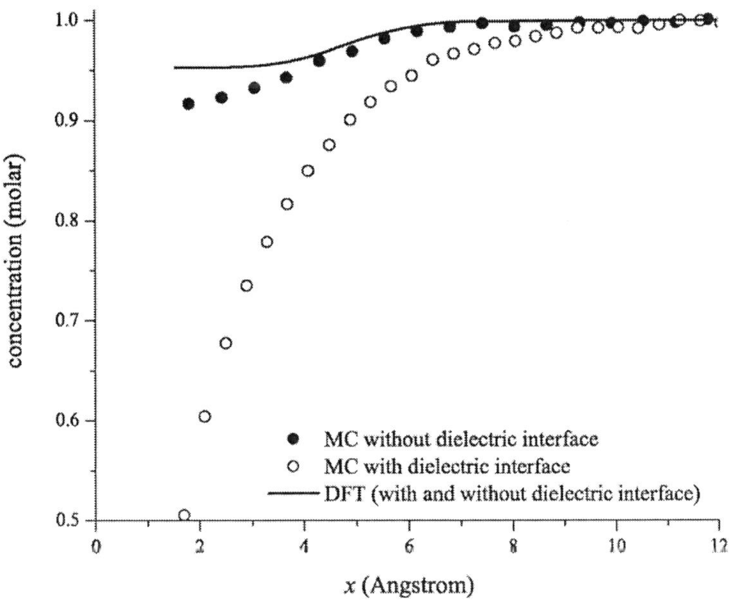


Figure 6. Ions in equilibrium near a smooth, hard, uncharged wall. All ions are the same size (3Å in diameter) and have charge +1 or -1. In this case, the cation and anion profiles coincide. The ion concentration profiles are different when there is a low dielectric behind the wall (open symbols) or not (solid symbols). The line is the result of density functional theory calculations. With and without a dielectric interface, the results for these DFT calculations are same for reasons discussed in the text.

Now let us consider the steady-state *mean* electrostatic potential $\phi(\mathbf{x})$. To describe the ion concentration profiles $p(\mathbf{x})$ and $n(\mathbf{x})$ in the steady state, ensemble averages of ion snapshots must be performed. When there are no dielectric interfaces (that is, $\varepsilon(\mathbf{x}) = \varepsilon$), $\phi(\mathbf{x})$ is given by

$$-\varepsilon \varepsilon_0 \nabla^2 \phi(\mathbf{x}) = e\left( p(\mathbf{x}) - n(\mathbf{x}) \right). \tag{64}$$

However, when there are dielectric interfaces, it is *not* true that ion distributions are governed by a mean potential given by

$$-\varepsilon_0 \nabla \cdot \left(\varepsilon(\mathbf{x}) \nabla \phi(\mathbf{x})\right) = e\left(p(\mathbf{x}) - n(\mathbf{x})\right). \tag{65}$$

This can be seen when monovalent hard-sphere ions (with all the ions being the same size) are in equilibrium near a smooth, hard, uncharged wall. The ion concentration profiles are shown in the **Fig. 6** when there is no dielectric interface (solid symbols) and when the dielectric coefficient behind the wall where the ions cannot go ( $x < 0$ ) is 2 (open symbols). In both cases, the dielectric coefficient is 80 for $x > 0$ (where the ions are) and the results were calculated with Monte Carlo simulations [4]. The simulations confirm the intuitive result that the low dielectric behind the wall repels ions from the wall.

In the **Fig. 6** it can be seen that the equilibrium (steady-state) cation and anion concentration profiles are identical; that is, at all locations **x**,

$$p(\mathbf{x}) = n(\mathbf{x}). \tag{66}$$

Substituting this fact into Eq. (65) gives that the right-hand side is identically zero. Consequently, the (unique) solution to Eq. (65) that satisfies the boundary condition that the potential is zero (grounded) somewhere is

$$\phi(\mathbf{x}) \equiv 0 \tag{67}$$

*regardless of the dielectric profile* $\varepsilon(\mathbf{x})$. In other words, for this case, having an inhomogeneous dielectric profile does not affect the mean electrostatic potential profile which is always identically zero even though the ions are still interacting with each other. Therefore, any theory that uses only the mean electrostatic potential to compute the steady-state ion concentration profile is doomed to failure because it will predict that the dielectric interface has no effect on the ion concentration profiles. The predictions of a density functional theory [7] (a theory more accurate than Poisson-Boltzmann) that in the case of equal-size ions uses only the mean electrostatic potential to compute the equilibrium ion concentration profiles confirms this (**Fig. 6**, line).

The problem described above has been discussed in the context of ion channels and possible fixes have been proposed [9; 14], but properly including the discreteness of charge near dielectrics in any statistical theory is still an open question.

## Conclusion

Electrostatics, in principle, seems simple. In practice, however, it is far from simple. One must be careful about the time and length scales of the system—and how they compare to the *desired* time and length scales, which can be orders of magnitude different. One must be careful about the approximations made to bridge those scales by modeling the relationship between charge distributions and the electrical potential. One must be careful about the boundary conditions of the system (either explicitly stated or implicitly assumed) that can dramatically change the computed potential. Since charges are ubiquitous in biology and electrical forces are very strong, an accurate physical understanding of a biophysical system requires an accurate and complete understanding of electrostatics.

# References

[1] Alberts, B; Johnson, A; Lewis, J; Raff, M; Roberts, K; Walter, P. *Molecular Biology of the Cell.* 4th ed. New York: Garland Publishing, 2002.

[2] Barthel, JMG; Buchner, R; Münsterer, M. Electrolyte Data Collection, Part 2a: *Dielectric Properties of Nonaqueous Electrolyte Solutions.* Kreysa G, editor. Frankfurt: DECHEMA, 1996.

[3] Berry, SR; Rice, SA; Ross, J. Physical Chemistry. 2nd ed. New York: Oxford, 2000.

[4] Boda, D; Gillespie, D; Nonner, W; Henderson, D; Eisenberg, B. Computing induced charges in inhomogeneous dielectric media: Application in a Monte Carlo simulation of complex ionic systems. *Physical Review E*, 2004, 69, 046702.

[5] Boyer, PD; Lardy, HA; Phillips, PH. The Role of Potassium in Muscle Phosphorylations. *Journal of Biological Chemistry*, 1942, 146, 673-682.

[6] Bracewell, RN. The Fourier Transform and Its Applications. 3rd ed. Boston: McGraw-Hill, 2000.

[7] Gillespie, D; Valiskó, M; Boda, D. Density functional theory of the electrical double layer: the RFD functional. *Journal of Physics: Condensed Matter*, 2005, 17, 6609-6626.

[8] Gilson, MK. Introduction to contiuum electrostatics, with molecular applications. 2000. http://www.biophysics.org/education/gilson.pdf.

[9] Graf, P; Kurnikova, MG; Coalson, RD; Nitzan, A. Comparison of Dynamic Lattice Monte Carlo Simulations and the Dielectric Self-Energy Poisson-Nernst-Planck Continuum Theory for Model Ion Channels. *Journal of Physical Chemistry B*, 2004, 108, 2006-2015.

[10] Griffiths, DJ. Introduction to Electrodynamics. 3rd ed. Upper Saddle River, NJ: Prentice Hall, 1999.

[11] Jackson, JD. Classical Electrodynamics. 3rd ed. New York: John Wiley and Sons, Inc., 1999.

[12] Marsden, JE; Tromba, AJ. Vector Calculus. 4th ed. New York: W. H. Freeman and Company, 1996.

[13] Saslow, W. Electricity, Magnetism, and Light. Toronto: Thomson Learning, Inc., 2002.

[14] Schuss, Z; Nadler, B; Eisenberg, B. Derivation of Poisson and Nernst-Planck equations in a bath and channel from a molecular model. *Physical Review E*, 2001, 64, 036116.

[15] Stryer, L. Biochemistry. 4th ed. New York: W.H. Freeman and Company, 1995.

[16] Wells, CM; Di Cera, E. Thrombin is a sodium ion activated enzyme. *Biochemistry*, 1992, 31, 11721-11730.

# INDEX

## A

## B



## C

## D

## E



## F



## G

## H



## I

## N



## O



## P

## T